Manuel Kalelessa Pina

Teaching strategy for introducing experimental activities

Manuel Kalelessa Pina

Teaching strategy for introducing experimental activities

Lê Chatelier Principle

ScienciaScripts

Imprint
Any brand names and product names mentioned in this book are subject to trademark, brand or patent protection and are trademarks or registered trademarks of their respective holders. The use of brand names, product names, common names, trade names, product descriptions etc. even without a particular marking in this work is in no way to be construed to mean that such names may be regarded as unrestricted in respect of trademark and brand protection legislation and could thus be used by anyone.

Cover image: www.ingimage.com

This book is a translation from the original published under ISBN 978-620-6-76126-6.

Publisher:
Sciencia Scripts
is a trademark of
Dodo Books Indian Ocean Ltd. and OmniScriptum S.R.L publishing group

120 High Road, East Finchley, London, N2 9ED, United Kingdom
Str. Armeneasca 28/1, office 1, Chisinau MD-2012, Republic of Moldova, Europe
Printed at: see last page
ISBN: 978-620-4-23545-5

Dedication

This work is dedicated to my father (Manuel Pina).

Thanks

First of all, thank God, the Almighty.

To my family for their attention during my training.

To my advisor, **Prof. Dr. José António Fins**, for accepting the invitation to be my advisor and for his unstinting efforts, which helped me to complete this work.

To the management, teachers and students of the 10th Class of the Physical and Biological Sciences Course (CFB) at the Joaquim Kapango-Huambo High School, for their attention and availability of time.

To the teachers and colleagues of the 4th edition master's degree 2021/2023. I would particularly like to thank Dr. Paulino Quintas, Dr. Justino Chipalavela, Dr. Petaxi Baptista and Dr. João Chefe for their support.

Manuel K. Pina

Summary

Research has shown that the subject of chemical equilibrium is extremely difficult for students, especially in Grade 10, where they begin to develop more complex concepts of the subject. The highly abstract nature of the content taught in the classroom has caused great difficulty in understanding chemical phenomena. The aim of this research is to develop a didactic strategy for the introduction of experimental activities in the treatment of the Lê Chatelier principle content at the Joaquim Kapango-Huambo High School. The previous study carried out through diagnosis and treatment revealed shortcomings, such as limited reference material for the practice of experimental activities, poor methodological orientation in the syllabus, lack of experimental activities, particularly in the content of Lê Chatelier's principle, among others. In order to minimize this problem, a didactic strategy was devised, which will enable chemistry teachers to find the motivation needed to achieve the objectives, thus promoting meaningful learning for the students. The methodology adopted consisted of a mixed approach, with a qualitative approach used to observe classes and interview the chemistry coordinator, followed by a quantitative approach that was carried out through surveys of teachers and students. The results obtained were processed by means of statistics using *SPSS software* and validated using the Mann-Whitney comparative method. The results of applying the strategy showed that the didactic strategy developed can significantly improve the teaching-learning process of the principle.

Keywords: teaching-learning process; didactic strategy; experimental activities; Lê Chatelier principle.

Contents

INTRODUCTION

Introduction

Chemistry is a well-known experimental science, but teachers' lack of mastery or adequate space for experimental activities are relegated to the background. The inclusion of experimentation in chemistry teaching is justified by the importance of its research role, helping students to understand phenomena and build concepts.

However, despite the encouragement to experiment, chemistry teaching is still overly theoretical. The use of experimental activities is an important tool for arousing students' interest in the phenomena being demonstrated and some of these activities do not require special facilities and can be carried out in regular classrooms when there are structural problems, such as the lack of a laboratory in the school. According to Silva (2021)the teaching of chemistry is becoming a major problem these days.

Chemical equilibrium is one of the most difficult and demanding subjects for secondary school students due to its complexity. However, it is vital for studying other aspects of the chemistry teaching-learning process. To this end, it is necessary to pay attention to learning difficulties and conceptual errors, as well as their possible origins (Quílez, Solaz, Castelló, & Sanjo, 1993). Some literature, such as that by Mouras, da Silva and Teixeira (2010)associate difficulties and conceptual errors on the subject of chemical equilibrium, in particular the Lê Chatelier principle, with the lack of experimental activities.

Given the opinions of some scholars, the author shares the idea of Freire et al. (2011)that each strategy has its advantages and disadvantages. There are also limitations in terms of feasibility on the part of students and teachers, and limitations in terms of promoting learning and developing students' skills. The integration of problem solving into experimental activities as a strategy goes beyond the traditional viewpoint of chemistry teaching and has a more effective potential in the teaching process.

In the Chemistry syllabus for Grade 10 of the Physical and Biological Sciences (CFB) course, one of the specific objectives for the subject of chemical equilibrium requires students to be able to identify variables that disturb chemical equilibrium and assess the consequences of altering the dynamics of an equilibrium system (INIDE & MED, 2020).

To achieve these ends, the Lê Chatelier principle is the main tool used in teaching, playing a fundamental role in the qualitative prediction of the evolution of a system in equilibrium based on changes in its properties. Studies by Canzian and Maximiano (2009) point to the Lê Chatelier principle as a qualitative rule for predicting possible changes in a system in chemical equilibrium.

The teaching-learning process of the Lê Chatelier principle is part of the Chemistry syllabus for the 10th class of the CFB course, the contents of which are only covered theoretically by the teachers due to a lack of mastery in carrying out experimental activities, which is a problem in the teaching-learning process of this principle.

In 1888, Lê Chatelier considered the principle to be purely experimental (Quílez-Pardo, 1995 as cited in Canzian and Maximiano, 2009, p. 2). In this sense, shortcomings on the part of students in understanding the principle are due to the lack of experimental activities. Research has shown that experimental activities can represent scenarios that prioritize students' emotional aspects, giving them the opportunity to learn concepts. Some literature states that students participating in such training are more interested in finding deeper explanations and meanings for the phenomena being demonstrated.

The inclusion of experimental activities in the treatment of the Lê Chatelier principle content has been advocated by several authors, such as Mouras, da Silva and Teixeira (2010), Junior (2017), Figuerêdo (2018) and others. These approaches aim to enhance knowledge of chemistry.

According to Saad (2005), carrying out experimental activities usually awakens in students a greater interest in the study of natural sciences. It is important to to explore/understand phenomena or scientific principles. Thus, the teaching of chemistry faces a number of challenges, including the lack of connection between the content taught in the classroom and the everyday occurrences of the students. (Sales & Silva, 2010). The lack of this connection has created a lack of interest and distance between these subjects and the students. Along with this obstacle, there is also a lack of experimental activity.

One of the fundamental aims of secondary education in Angola is to provide students with learning that has meaning for their lives. Chemistry teaching for the 2nd Cycle should enable students to understand both the chemical processes themselves and the construction of scientific knowledge, which is closely linked to technological, economic, social, political and environmental applications, in order to meet the demands of contemporary society. (INIDE, 2013). It is therefore necessary to change teaching strategies in order to provide meaningful learning for students.

According to Marcondes (2006)experimentation in the teaching-learning process of chemistry has its importance justified when one considers its pedagogical function of helping students to understand chemical phenomena and concepts. The clear need for students to relate to the phenomena to which the concepts refer justifies experimentation as part of the school context, to provide a link between theory and practice.

Several researchers have studied methodological alternatives to contribute to the quality of teaching, and one of the ways pointed out in the literature to improve learning in chemistry is the use of experimental activities as one of the possibilities for working on the concepts of the subject in the classroom. Most of the literature has been more concerned with the theoretical aspects of the approach to chemical equilibrium content, in particular the Lê Chatelier principle, thus leaving shortcomings in the practical aspects.

The recognition that experimental activities contribute to the teaching-learning process has become a real fact among science teachers, particularly in chemistry. However, there are cases where even simple activities are not carried out due to a lack of materials and conditions.

In the exploratory stage of the research, the questioning of the aspects related to the fulfillment of experimental activities in the teaching-learning process of Chemistry, in particular in the treatment of the contents of the Lê Chatelier principle, at the Joaquim Kapango-Huambo high school, made it possible to identify weaknesses that are evident in:

- Limited reference material for experimental activities;

- Lack of experimental activities due to a lack of means to develop them in the chemistry teaching-learning process;

- Insufficient methodological guidelines in the Chemistry Grade 10 syllabus and student handbook;

- Students' lack of interest in chemistry,

- There is little creativity on the part of teachers when it comes to designing and implementing experimental activities related to the contents of the Lê Chatelier principle, which to a certain extent does not contribute to meaningful learning and student motivation.

In this context, there is a contradiction between the need for students to interpret the Lê Chatelier principle and the limitations of the teaching-learning process, which does not go beyond experimental activity as a prerequisite for effective learning aimed at solving community problems.

According to Law no. 32/20 in its article no. 33, one of the specific objectives of the 2nd cycle of General Secondary Education is:

> "To offer complete and in-depth training in a given area of knowledge [...]; to prepare the student so that immediately after completing the cycle they are qualified and can immediately enter higher education institutions; to develop practical experience, strengthening the mechanisms of convergence between the school and the community, strengthening the innovative and intermediary role of the school" (Law no. 32/20 of August 12 amending Law no. 17/16, p. 4425).

However, the recognition that experimental activities provide students with more meaningful learning is a real fact. According to Gonçalves (2005)experimentation is the starting point of scientific endeavor and the discovery of relationships, the identification and classification of functional relationships between phenomena.

In view of the above, the following research problem was formulated: shortcomings in the teaching-learning process of the Lê Chatelier principle, which limit the effective learning of students in the 10th grade of the CFB course at Liceu Joaquim Kapango-Huambo because it does not go beyond experimental activities.

These elements contribute to the determination of the research object: the teaching-learning process of Chemistry in the 2nd Cycle of General Secondary Education.

The scope of the research was set **at** experimental activities in the treatment of the Lê Chatelier principle in the 10th grade of the Physical and Biological Sciences course.

The aim of the research was **to** develop and implement a didactic strategy using experimental activities to deal with the content of the Lê Chatelier principle, as a prerequisite for effective student learning.

The research follows a quasi-experimental exploratory-descriptive design with the following hypothesis: the implementation of a didactic strategy based on experimental activities in the treatment of the Lê Chatelier principle content improves student learning.

In order to verify the advanced hypothesis, the following variables were manipulated:

Independent variable: teaching strategy.
Dependent variable: student learning.
In order to achieve the objective, the following research tasks were carried out:
1. Systematization of the theoretical foundations of the chemistry teaching-learning process and experimental activities;
2. Characterization of the current state of the teaching-learning process of the Lê Chatelier principle and that of experimental activities;
3. Elaboration of the model and didactic strategy for the introduction of experimental activities in the treatment of the Lê Chatelier principle content in the 10th grade of the Physical and Biological Sciences course at the Joaquim Kapango-Huambo High School;
4. Implementation of the didactic strategy in the teaching-learning process;
5. Validation of the teaching strategy using the Mann-Whitney comparative method.

The following research methods were used to carry out the tasks:

Theoretical methods

Historical-logical: used to characterize the background and essential features of the historical evolution of the object of research and field of action.

Synthesis analysis: used throughout the research process, particularly to determine the theoretical foundations of the proposal, as well as the empirical results obtained in drawing up conclusions and suggestions;

Induction-deduction: to integrate the general and the particular in the analysis of the theoretical concepts that underpin the research, as well as in the study of particular cases that allow conclusions and generalizations to be reached.

Modeling: in the development of the model;

Systemic-structural-functional: in the development of the model and didactic strategy and the work as a whole.

Empirical methods and techniques

Observation: allowed observation of chemistry classes to see how teachers approach chemical equilibrium content, in particular the Lê Chatelier principle.

Interviews: used to record information from management and teachers.

Questionnaire survey: used to gather information from teachers and students according to the indicators of the problem raised.

Documentary analysis: used to analyze the programs, the curriculum of the 2nd Cycle of Secondary Education, the Basic Law of the Education and Teaching System, as well as other essential documents in the teaching-learning process of Chemistry.

Pre-test and post-test: applied to students in the 10th class of the CFB course, with the aim of evaluating the effectiveness and relevance of the didactic strategy in the chemistry teaching-learning process.

Experimentation: to implement the proposal.

Statistical methods

Descriptive statistics: for processing and analyzing the data collected during the implementation of empirical methods.

Inferential statistics (Mann-Whitney comparative method): to validate the teaching strategy.

Population and sample

The research involved a population of 61 individuals, made up of 1 Chemistry course coordinator, 5 teachers and 55 students representing two classes (10.1

and 10.4a) in the 10th Class of Physical and Biological Sciences at Liceu Comandante Joaquim Kapango - Huambo. The sample was taken intentionally and consisted of 1 coordinator, 4 teachers and 47 students: 24 students from class 10.1 (experimental group), 23 from class 10.4a (control group).

The relevance of the research lies in the application of the didactic strategy that mobilizes the active participation of students in the different phases, promoting motivation, effective learning and a focus on solving community problems. Furthermore, the topic is pertinent and topical because it is framed within the context of the Angolan Ministry of Education's educational policy, which involves training students who are the protagonists of their own learning.

The theoretical contribution is expressed by the model as a theoretical construct that supports the didactic strategy based on experimental activities in the treatment of the content of the Lê Chatelier principle. The practical contribution is expressed in the didactic strategy as a tool for teachers' pedagogical practice, an instrument for effective student learning.

This dissertation is structured as follows: introduction, which presents the theoretical-methodological design of the research; first chapter, which refers to the theoretical foundations related to the subject under study and the characterization of the teaching-learning process; second chapter, which presents the model and didactic strategy, and third chapter, which presents the validation of the didactic strategy.

The work also includes conclusions, recommendations, bibliographical references, appendices and annexes.

CHAPTER I. THEORETICAL FOUNDATIONS OF THE RESEARCH

Chapter I. Theoretical foundations of the research

This chapter deals with the theoretical foundations that underpin the teaching-learning process of chemistry, based on treatments of concepts and analysis of the current situation that have repercussions within this process, specifically in experimental activities.

1.1. Systematization of theoretical foundations

1.2. Conceptions of the chemistry teaching-learning process and the use of experimental activities

The teaching-learning process is made up of four elements: the teacher, the student, the content and the environmental variables (characteristics of the school), (Moreira , 1986). According to the theory of (Piaget, 1969 as cited in Wongombo, 2015, p. 12), thought is the basis on which learning is built. It is a way of manifesting intelligence. It is assumed to be a biological phenomenon, due to the neural basis of the brain and the whole body, subject to the body's maturation process.

Learning, according to Vygotsky's research, defined by (Oliveira, 1993 as cited in Ogasawara, 2009), is "the process by which the subject acquires information, skills, attitudes, values, etc. from their contact with reality, the environment and other people". Two types of learning have been distinguished, namely casual and organized learning. Casual learning is almost always spontaneous, arising naturally as a result of interaction between people and with the environment in which they live. Organized learning, on the other hand, is something that has the specific objective of mastering certain knowledge, skills and norms of social coexistence (Libaneo, 2006, p. 82).

Teaching is a systematic response to the natural needs of the educational process. It should stimulate the production of knowledge and the development of a knowledgeable, reflective and active subject in the learning process (Behrens, 2005). Teaching as a process is an intermediary activity, through which conditions and means are created for students to become active subjects in the assimilation of knowledge. (Libaneo, 2006, p. 89).

Thus, the process of teaching science, particularly chemistry, requires the use of teaching methods that enable students to acquire scientific and technical knowledge for the benefit of society. In this context, Rodrigues (2015)state that chemistry is a subject that requires teachers to undergo continuous and ongoing professional development.

Chemistry is part of the school curriculum in the 1st cycle, 2nd cycle and in higher education. The failure of experimental activities in the chemistry teaching-learning process is a fact. The reality shows that the teaching of chemistry is not at a low level, it simply doesn't keep up with the evolution that is currently taking place, because in practical terms, its methodology is carried out exclusively in a theoretical way, the study of the subject is understood only as a process of accumulating knowledge. Sousa (2015)states that concepts, laws and principles are taught in an abstract way, far removed from the daily lives of students and teachers.

According to Lima (2012)it is essential that chemical knowledge is presented to students in such a way that they can interact actively and deeply with their environment, realizing that this is part of the world in which they are also actors and responsible. The construction of this knowledge should include activities of recognition, identification, manipulation, differentiation, characterization, interpretation, application, problem-solving, discussion and explanation.

The student's active and creative participation in the teaching-learning process allows them to bring modern reality and social practice closer together, making the school a place for exchanging ideas and conceptions, a place where everyday knowledge is transferred, becoming scientific, and the student develops in an integral way, which has always been present in pedagogy since ancient times. Despite some cases, his concept falls within traditional pedagogy, but his ideas already show important signs of forming a transformative and active position in the student (João, 2007).

1.2.1. Experimental activities in the teaching and learning of chemical equilibrium, particularly in the study of the Lê Chatelier principle

Experimentation is, in fact, a way of better understanding and mastering chemical concepts. This means that it is extremely important to use experiments in chemistry teaching, given their pedagogical nature in helping students to understand the phenomena to which these concepts refer (Nerci, 1989).

Research in science teaching has revealed important evidence that the majority of chemistry teachers believe in experimental activity as a way of providing students with solid knowledge (Hodson , 1994).

Experimental activities can be used for a variety of purposes and make varied and important contributions to science teaching and learning. According to Carvalho et al. (2005)in addition to facts, other types of content or knowledge (conceptual, procedural, attitudinal) can be processed or favored in experimental activities. From this point of view, the following are some possible contributions that experimental activities can make to science teaching and learning: (i) motivate and arouse students' attention; (ii) develop the ability to work in groups; (iii) develop personal initiative and decision-making; (iv) stimulate creativity; (v) improve the ability to observe and record information; (vi) allow learning to analyze data and propose hypotheses for phenomena; (vii) allow the study of scientific concepts; (viii) allow detecting and correcting students' conceptual errors; (ix) allow understanding of the relationship between science, technology and society, and others. (Numbi , 2016).

Tomalela (1998)considers experimentation in chemistry teaching to be a means of providing meaningful learning for students, since every experience contains content.

Some authors see experimentation as the realization of phenomena that allow students to establish causal relationships that support their ideas and in one way or another influence their attitudes towards learning (Moreira, 1980; Ndala, 2007 as cited in Catumbela, 2016, p. 22).

According to Marcondes (2006), the importance of experimental activities in chemistry teaching in the learning process is justified if we take into account their pedagogical function of helping students to understand chemical phenomena and concepts. The obvious need for students to relate to the phenomena to which the concepts refer justifies experimentation in the school context without a gap between theory and practice.

According to Araujo and Abib (2003)experimental activities can be classified into three types of approaches or modalities, which are: demonstration, verification and investigation activities, which are characterized as follows:

Demonstrative experimental activities are those in which the teacher carries out the experiment and the students just observe the phenomena taking place, i.e. the students' role is to observe the experiment and, in some cases, offer explanations;

Experimental verification activities are activities carried out with the aim of testing or confirming a law or theory, i.e. the students' role is to carry out an experiment and explain the phenomena observed;

Experimental research activities are those strategies that allow students to play a more active role in the process of constructing knowledge, and the teacher to become a facilitator or mediator of this process. In other words, the students' role is to explore, plan and carry out activities and discuss explanations. At the heart of inquiry and experimentation is its ability to ensure greater student participation in all stages of the investigation, from the interpretation of the problem to its possible solution.

Many of the traditional approaches to experimentation (demonstration, verification) offer students little opportunity to analyze problem situations, collect data, develop and test hypotheses, argue and discuss with colleagues (Suart and Marcondes, 2008 as cited in Oliveira, 2010).

Studies by Ferreira, Romeu and Dácio (1997); Mouras, da Silva, and Teixeira (2010)have shown that the use of experiments in this principle contributes to the construction of scientific concepts by students. It is therefore necessary to carry out experimental activities in Angolan schools in order to improve the chemistry teaching-learning process.

1.3. Contents chemical equilibrium as a learning topic for grade 10

The subject of chemical equilibrium is considered extremely important for understanding everyday facts and phenomena from the point of view of science. It is from the 10th grade onwards that a more in-depth approach to the subject begins. The daily use of chemical equilibrium according to Figuerêdo (2018)can be observed in cellular respiration through the transport of O_2 to the cells of all organisms. After respiration, oxygen comes into contact with hemoglobin (Heme) in the blood, giving rise to oxyhemoglobin (HemO$_2$). Drinking soft drinks can increase the concentration of acid in the stomach, or reduce the pressure, or even increase the temperature, limited by the solubility of CO_2 in water, which decreases with increasing temperature. In tooth enamel through hydroxyapatite (Ca_5 OH(PO)$_{43}$), which under the influence of pH can cause demineralization or mineralization of the tooth. This subject is one of the most important and complex in the teaching of chemistry in secondary schools (Quílez & Solaz, 1995).

This subject is highly conceptual and requires other knowledge to understand, such as chemical reactions, gases, stoichiometry, kinetic concepts and thermochemistry. Thus, teaching this content is a favorable circumstance for integrating or applying previous concepts and for diagnosing remaining difficulties in order to overcome them with appropriate strategies that facilitate learning.

According to Farias (2017)According to Farias (2017), the concept of chemical equilibrium has great potential and implications for chemistry teaching, since the scope of the subject is broad and it is used in the main areas of chemistry such as organic and inorganic chemistry, etc.

The studies by Canzian and Maximiano (2010); Lima Verde (2019) and Silva (2021)state that chemical equilibrium is one of the most difficult subjects to teach and learn. Most chemistry teachers consider chemical equilibrium to be one of the most difficult subjects to teach (Finley, Stewart, & Yarroch, 1982)and students also consider it one of the most difficult to learn (Butts & Smith, 1987). Some authors offer various didactic approaches to facilitate the teaching-learning process of this content, such as computer simulations, analogies and experiments.

Mouras et al. (2010); Bedin and Cassol (2016)propose the use of analog prototypes and experimental activities to make the content of chemical equilibrium, in particular the Lê Chatelier principle, clear and understandable to students. These methods can minimize learning difficulties. Among the difficulties that students face in learning this content, understanding, correct application and use of the Lê Chatelier principle occupy a prominent place. (Queliz, 1995-1998). A number of secondary school students and teachers have difficulties in applying Lê Chatelier's rules to decision problems.

1.4. Historical background to the Lê Chatelier principle

The French chemist and engineer Henri Louis Lê Chatelier formulated Lê Chatelier's principle in 1884, based on the work of J. G. Van't Hoff and G. Lippmann. Lippmann. The apparent simplicity of the principle favored its application in numerous works that had great repercussions in the 19th century and were dedicated to those that used mathematical treatments based on the principles of thermodynamics (Canzian & Maximiano, 2010).

This principle was formulated by Lê Chatelier to describe reversible chemical reactions, in which an increase in the concentration of one of the initial substances leads to a shift in the equilibrium towards the formation of reaction products (Gurov et al. 2004, as cited in Bauman, 2015, p. 5). This principle was later generalized by Braun to equilibrium thermodynamic systems.An example of the application of Lê Chatelier's principle in mechanics is the gyroscopic effect, which consists of the tendency of the gyroscope, when an external moment of forces is applied to it, to rotate its axis in such a way as to reduce the angle between the moment vector of the gyroscope and the moment vector of the forces (Frish and Timoreva, 2006, as cited in Bauman, 2015, p. 5).

By way of example, it follows that under external influences on the mechanical system such changes occur that they tend to reduce the result of these influences. When any state of a system in equilibrium changes, the system tends to seek a new state that minimizes the imposed changes until it reaches a new state of equilibrium. This change in initial conditions will cease when the speeds of the two reactions equalize and the concentrations of reactants and products remain constant again (Santos, 2016 and Mol, 2017).

Referring to the use of this principle in the 2nd cycle of secondary education, in particular in the 10th grade, the focus of this study, some research in the 1990s continued to address other situations in which it is limited and can generate conceptual errors in these cases. The lack of experimental activities has contributed significantly to these conceptual errors. In 1887, the German chemist F. Braun published an article in which he studied the effect of pressure on the solubility of salts in water (Braun, 1887, quoted in Quílez-Pardo, 1995). This work led him to further establish the effects of variable changes in thermodynamic systems (Braun, 1887, Lê Chatelier 1888, as cited in Quílez-Pardo, 1995). These contributions led German authors to call it the "Lê Chatelier-Braun principle".

According to Lê Chatelier (1888), as quoted in Quílez-Pardo (1995), this principle is purely experimental and he established various factors for analyzing equilibrium: temperature, pressure, condensation (concentration) and so on, proposing a specific statement for each.

1.4.1. The teaching-learning process of the Lê Chatelier principle and applications practices

According to Lê Chatelier's principle, when any state of a system at equilibrium changes, the system tends to seek a new state that minimizes the imposed changes until it reaches a new state of equilibrium. This change in initial conditions will stop when the speeds of the two reactions become equal and the concentrations of the reactants and products remain constant again (Santos and Mol, 2016).

The principle addresses how a reaction will try to adjust the quantities of reactants and products until equilibrium is restored again, i.e. so that the quotient of the reaction Q becomes equal to the equilibrium constant K. This may sound very theoretical, but Lê Chatelier's principle is applicable in many situations during a chemical reaction.

For nomenclature purposes, the equilibrium shifts to the right when the change introduced favors the formation of products. On the other hand, when a change favors the formation of reactants, the equilibrium is said to shift to the left (Santos and Mol, 2016).

The INIDE syllabus for general secondary education highlights the terms "identify", "recognize" and "predict" factors that affect equilibrium through the Lê Chatelier principle. (INIDE & MED, 2020) .

Canzian and Maximiano (2010) emphasize that the Lê Chatelier principle has limitations for indiscriminate use. And it presupposes that teachers will be transparent about the limitations, so that its only application is to understand the dynamic nature of chemical equilibrium. To illustrate, we can give the reaction of ammonia formation $N_{2(g)}$ + 3 $H_{2(g)} \square$ $2NH_{3(g)}$, which, when gaseous nitrogen is added to the mixture if a gas is in equilibrium, Lê Chatelier's principle ensures that there will be a shift in the equilibrium state towards producing more ammonia, since a system in equilibrium must consume nitrogen, producing more ammonia. This effect would not only occur at constant temperature and volume.

1.4.1.1.　　Practical application of Lê Chatelier's principle

This paper has listed three (3) practical applications of the Lê Chatelier principle in the interpretation of chemical equilibrium:

1.　　The effect of temperature on chemical equilibrium, an increase in temperature generally implies an increase in the speed of the reaction; its molecules, subjected to higher temperatures, are more energetic, i.e. they have greater kinetic energy. More collisions will probably result in more reactions. According to Santos (2016) and Mol (2017), the speed of direct and indirect reactions increases with increasing temperature.

ΔH is a heat variation quantity, which can indicate whether a reaction is energetically exothermic or endothermic. This can be exemplified when NO gas$_2$ is in equilibrium with N gas O_{24} . The dark brown NO_2 gas becomes colorless when we place the tube in a container with ice, which means that the temperature has favored the formation of N O_{24} gas, whose reaction is exothermic (Santos, 2016 and Mol, 2017).

2.　　The effect of concentration, when there is an increase in the concentration of reactants, leads to more reactive particles for collisions, causing an increase in the speed of the direct reaction. On the other hand, an increase in the amount of products of the reaction will cause an increase in the reverse reaction and so the system will continue to exist until a new state of equilibrium is reached, in which the speed of the direct reaction will equal the speed of the reverse reaction.

3.   The effect of pressure: an increase in the pressure of a gaseous system, under constant temperature conditions, implies an increase in the pressure of the gases present. If there are more gaseous reactant molecules than products, the pressure of the reactants will be greater in relation to the products, causing more collisions between the reactant molecules, favoring a direct reaction.

On the other hand, when the pressure decreases, the reverse reaction will be favored. Under equal conditions of temperature and pressure, gases occupy volumes proportional to the amount of substance. Thus, it can be argued that when the pressure increases, the equilibrium shifts the reaction to a smaller volume and when the pressure decreases, it shifts to a larger volume. Changing any of the conditions can affect the chemical equilibrium state of the system, causing a reaction in such a direction that it tends to produce a change in the opposite direction to the change in the external condition.

This principle is not limited to chemistry, but also applies to other areas such as biology, physics, industry and others. In biospheric processes, it makes it possible to predict the development of the ecological situation based on the description of the biosphere as a self-regulating system. (Bauman, 2015).

In physics, for example, the principle can be applied to a thermodynamic system that goes out of equilibrium not only due to external influences, but also under the influence of strong fluctuations that arise in the system itself. In this case, such changes occur in the system that it tends to return to the equilibrium state as quickly as possible, while the entropy of the system increases.

Therefore, the use of experimental activities in the treatment of the principle, in addition to motivating students, will provide more lasting learning in the understanding of physical and chemical processes.

1.5. Characterization of the current state of the chemistry teaching-learning process and experimental activities at Liceu Joaquim Kapango-Huambo

In order to characterize the initial state of the teaching-learning process of the Lê Chatelier principle and experimental activities, questionnaire surveys were administered to the coordinator, teachers and students (Appendices I, II, III), with the following objectives:

- To identify the main difficulties in dealing with the content of the Lê Chatelier principle at Joaquim Kapango-Huambo High School.
- To determine the potential of the content of the Lê Chatelier principle in order to develop a didactic strategy for introducing experimental activities to deal with the principle.

The diagnoses made showed that experimental activities are not an effective practice among 2nd cycle chemistry teachers, where the traditional descriptive method prevails. According to Bedin and Cassol (2016), the great difficulty in understanding chemical concepts is due to their abstraction and, for this reason, they suggest using analog prototypes and experimental activities to make the content clear and understandable to the student.

In the process of teaching and learning chemistry, it is impossible to conduct a lesson using only a whiteboard and chalk as a teaching resource. Experiments play a fundamental role in chemistry teaching, and research has shown their importance in involving students in the investigation process, linking experimental activities with semi-open problem solving, which can be very effective in students learning concepts, procedures and attitudes (Goy & Santos, 2008). It is known that students find it difficult to learn content and therefore feel frustrated because they do not consider themselves capable of studying chemistry, and also because they do not realize the importance of this subject in their daily lives (Wongombo, 2015).

The reality of the current situation of the chemistry teaching-learning process at the Joaquim-Kapango High School, in the treatment of the subject of chemical equilibrium, in particular the Lê Chatelier principle, can be seen in the following situation extracted from the diagnostic results of the selected samples.

As for the students:

- Difficulties in understanding some aspects of the subject;

- They haven't had any experimental lessons;

- They find that experimental activities make their learning more motivating and meaningful;

- They consider it very important to implement a didactic strategy for introducing experimental activities into the content of Lê Chatelier's principle;

- Expressing the need for frequent experimental activities in chemistry classes.

As for the teachers:

- The teaching-learning process in chemistry is far from what is expected, despite the efforts made by the school management.

- They were unanimous in their response that students always feel the need to put theory into practice;

- They were unanimous in their response that experimental activities make a significant contribution to students' learning;

- As for the inadequacy of experimental activities, they say that this is not only expressed in the lack of materials and reagents, but also in the lack of guiding materials for experimental activities;

- The analysis of the syllabus and the 10th grade chemistry student handbook allowed us to determine the shortcomings of the methodological guidelines in the syllabus and the 10th grade chemistry student handbook, particularly with regard to the Lê Chatelier principle.

- The interview with the coordinator revealed the existence of a well-equipped laboratory that offers working conditions;

- Teachers have not carried out any experimental activities related to the subject.

A significant part of the difficulties in teaching chemistry lies in its experimental nature. Schools do not perceive experimental activities as a method of assessing and stimulating learning. In this work, the author observed that the inclusion of experimental activities increases interest and generates positive incentives in the classroom. Thus, a didactic strategy for introducing experimental activities for the treatment of the principle will contribute to improving teaching practice and meaningful student learning.

Conclusion of chapter I

The definition of a system for improving the teaching-learning process of chemistry through experimental activities will emphasize the need to obtain better results in this process, thus creating a better development of the logical abilities of Grade 10 students, in particular in the interpretation of Lê Chatelier's principle in its treatment through experimentation.

The current main trend in the education system is towards constructivism, placing a strong emphasis on the value of observation and direct experimentation, suggesting changes in cognitive structure in the students' learning process. In addition, carrying out experiments in the chemistry teaching-learning process is a method that guarantees the consolidation of students' knowledge, while at the same time independently applying this knowledge in other specific situations.

CHAPTER II. DIDACTIC STRATEGY FOR INTRODUCING EXPERIMENTAL ACTIVITIES IN THE TREATMENT OF THE LÊ CHATELIER PRINCIPLE CONTENT

Chapter II. Didactic strategy for introducing experimental activities in the treatment of the content of the le chatelier principle

This chapter presents the theoretical foundations that support the model and teaching strategy based on the introduction of experimental activities in the treatment of the Lê Chatelier principle content, with a focus on the development of skills and values in the teaching-learning process through a flexible, active and integrated set of actions, supported by the principles of teaching-learning that links theory with practice.

2.1. Results of the surveys of teachers, students and the coordinator of the Physical and Biological Sciences course

2.1.1. Analysis of the results of the teacher questionnaire.

In order to achieve the aims of this study, teachers and students were surveyed. A total of 5 teachers who teach 10th grade chemistry were surveyed, including the chemistry course coordinator, aged between 39 and 52. All the teachers surveyed are graduates, 90% of whom have a degree in biology teaching. The shortage of professionals with specific training in Chemistry highlights the need to increase reflection scenarios and address the training of Chemistry teachers in greater depth in order to make up for these shortcomings (Silva, 2011).

The teachers surveyed have extensive professional experience in teaching chemistry, with an average of 22 years. These results are pertinent, since time in the job contributes significantly to improving teaching methodologies.

With regard to experimental activities, the majority said that they had carried out experimental activities in chemistry lessons (see figure 1).

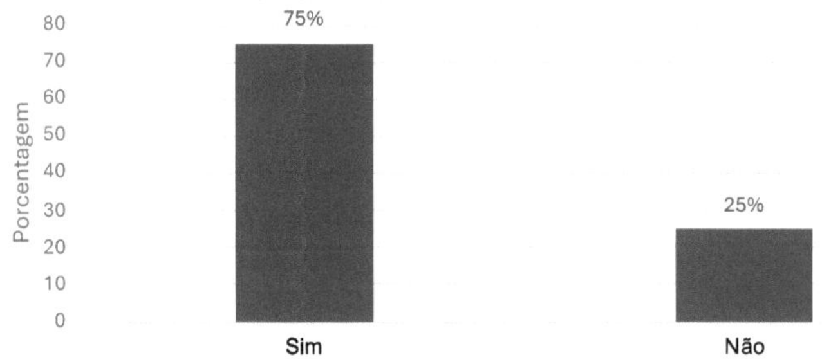

Figure 1Teachers' responses on the development of experimental activities in general

Of the 4 teachers surveyed, which corresponds to 100%, 75% said that they had carried out experimental activities in general during lessons. As for the frequency of activities per term, 75% of the respondents said that they had done them three times a term, which is the minimum acceptable by some authors.

When the teachers' assumptions were compared with the students' conceptions, more than 80% of the students said that they had never had any experimental activities.

With regard to the content of Lê Chatelier's principle, the majority of teachers said that they had carried out experimental activities when dealing with the content (see figure 2).

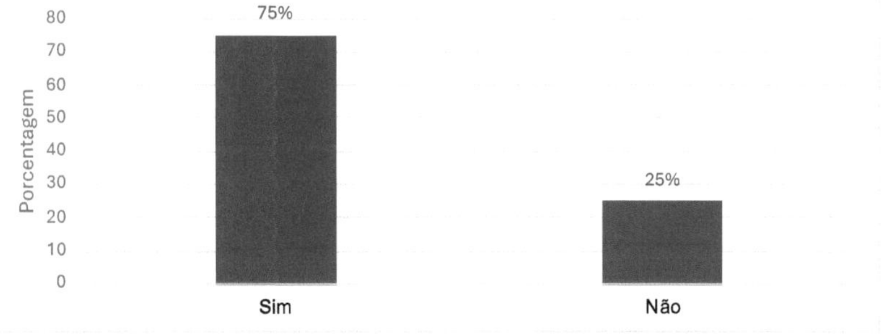

Figure 2Teachers' responses on the development of experimental activities in the treatment of the "Lê Chatelier principle" content.

Source: author

Figure 2 shows that 75% of the teachers surveyed have carried out experimental activities when dealing with the "Lê Chatelier principle" content during their lessons. These statements contradict the opinion given by the course coordinator, who stated that there were no experimental activities related to the Lê Chatelier principle, due to various factors. Furthermore, the students' opinion and the observations made also showed that they did not exist.

The lack of experimental activities in the classroom not only hinders learning, but also demotivates students. Teachers should look for ways to carry out experimental activities three times a term (once a month). Practical exercises can be considered an excellent strategy for creating contextualized problems and thus answering students' questions during interaction with this activity (Guimarães , 2009).

As for the importance of implementing the didactic strategy for introducing experimental activities in the treatment of the Lê Chatelier principle, the teachers were unanimous in stating that it is very important to implement it.

Experimental activities are important in the teaching of chemistry. As well as providing an environment conducive to meaningful learning, they also create interest and motivate students to develop their learning in a meaningful way.

Question about difficulties in carrying out experimental activities (table 1). In this question, values between 1-5 were assigned with different levels of agreement, being: 1 - I totally disagree; 2 - I don't agree; 3 - I'm undecided; 4 - I agree; 5 - I totally agree.

Table 1Teachers' responses on the difficulties in carrying out experimental activities in the "Lê Chatelier principle" treatment.

		Lack of materials and reagents	Lack of a document to guide experimental activities	Associate experiment with content	Selection of materials and reagents
NO.	Valid	4	4	4	4
	Omitted	0	0	0	0
Average		3.25	3.75	3.00	3.25
Median		3.50	4.00	3.00	3.50
Fashion		4	4	3	4
Minimum		2	3	2	2
Maximum		4	4	4	4

Source: author

As can be seen in table 1, teachers point to the "*lack of a document* to *guide experimental activities*" as one of the factors preventing them from being carried out. There is therefore a need to draw up a proposal that responds to this shortcoming.

Some authors believe that conducting experimental activities depends a lot on the teacher, because even with a lack of materials and reagents, most of the time it is necessary to use alternative or readily available materials. To this end, it is important that the teacher tries to overcome their own difficulties and the lack of materials in order to find alternatives to make learning more meaningful for the students. In this sense, the importance of the teacher's creativity should be emphasized.

The respondents pointed out a number of advantages that experimental activities can bring to the chemistry teaching-learning process. Their opinion, based on the pedagogical experience they have accumulated over the years working as chemistry teachers, is shown in Table 2 in more detail. In this question, values between 1-5 were assigned with different levels of agreement, being: 1 - I totally disagree; 2 - I don't agree; 3 - Undecided; 4 - I agree; 5 - I totally agree.

Table 2Teachers' responses on what experimental activities can provide.

		An environment conducive to more complete learning	More motivation to learn	More instructive teaching	Better understanding of the content	More meaningful learning	Students' handling skills in laboratory work
NO.	Valid	5	5	5	5	5	4
	Omitted	0	0	0	0	0	1
Average		4.60	5.00	4.40	4.80	5.00	5.00
Median		5.00	5.00	4.00	5.00	5.00	5.00
Fashion		5	5	4	5	5	5
Asymmetry		-.609		.609	-2.236		

A general analysis of the mean, median and mode of table 2 shows that teachers totally agree with the variables presented in the table. This demonstrates the importance of using experimental activities.

The question about the contribution of the experimental activity to building students' knowledge. In this question, values were assigned, where: 1- Rarely; 2- Always; 3-Never.

Table 3Teachers' opinion on the contribution of experimental activities to building students' knowledge.

		Frequency	Percentage	Valid percentage	Cumulative percentage
Valid	Always	4	100.0	100.0	100.0

Source: author

The teachers surveyed were unanimous in stating that experimental activities always contribute to the construction of knowledge, thus corroborating the ideas defended by various authors, such as Zanon and Silva (2000)who state that experimentation can contribute to meaningful learning, as long as a relationship is established between theoretical and practical knowledge.

2.1.2. Analysis of the results of the student questionnaire.

In part, as has already been shown in the sample, the questionnaire survey guide applied to the students was answered by a total of 47 students, 20 of whom were male and 27 female. The results and an analysis of each of the questions are presented below.

To find out how much you like the subject: values were given to reflect the degree of satisfaction with the subject, where: 1- I don't like it; 2 - A little 3 - I like it.

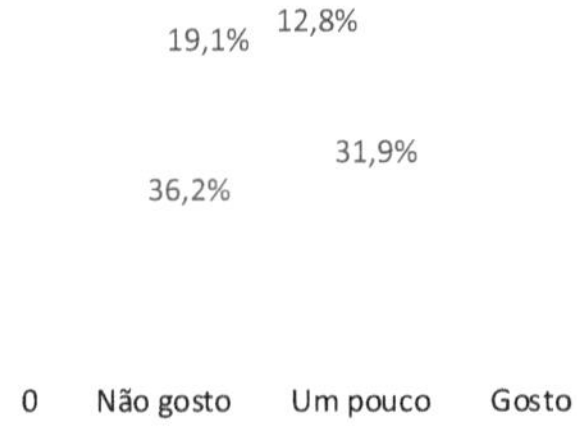

Figure 3Students' responses on their enjoyment of chemistry.

Source: author

Figure 4 shows that the majority of students don't like chemistry. As for their justifications, they didn't expect lessons associated with a lot of theory. This context leads us to agree with the studies by Lígia al. et. (n.d.), who state that if there is no link between theory and practice in teaching, the content will not be of great importance to the student's education or will contribute very little to their cognitive development. Thus, it seems that the teaching of chemistry has not created the conditions for students to understand the concepts in terms of their application in everyday life.

For the question about how they would like their chemistry lessons to be developed. In this question, values between 1-5 were assigned with different levels of agreement, being: 1 - I totally disagree; 2 - I don't agree; 3 - I'm undecided; 4 - I agree; 5 - I totally agree.

Table 4Students' opinions on how they would like their chemistry lessons to be conducted.

		Based on exercises (use of formulas)	No experimental activities included	With the inclusion of experimental activities	Lectures with illustrative slides
NO.	Valid	47	47	47	47
	Omitted	0	0	0	0
Average		2.45	1.96	3.62	3.36
Median		3.00	2.00	4.00	4.00
Fashion		1	1	5	5
Asymmetry		.021	.587	-1.239	-.886
Minimum		0	0	0	0
Maximum		5	5	5	5

Source: author

Looking at table 4 in general, by the minimum and maximum, some students gave no opinion and some totally agreed with the variables presented depending on the question asked. The variable *"no inclusion of experimental activities",* based on the average number of opinions, is more of an indecision. However, according to the mode, a large proportion totally disagree that lessons are given without the inclusion of experimental activities. The variable *"with* the *inclusion of experimental activities"* shows that the majority agree that chemistry lessons should *include experimental activities.* The asymmetry is negative. More details can be found in Figures 5 and 6.

Figure 4Students' opinion on the development of chemistry lessons without the inclusion of experimental activities.

Source: author

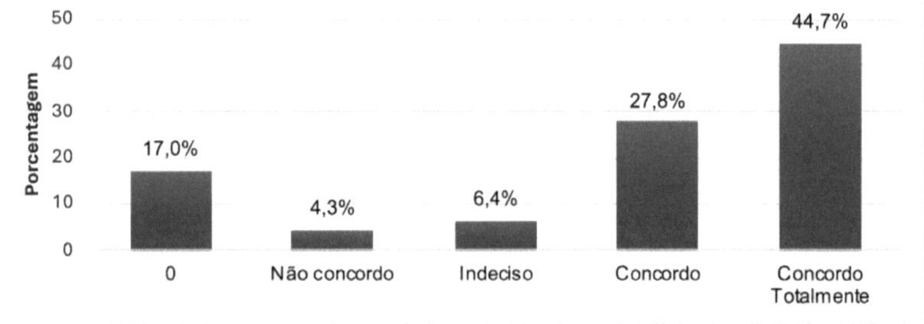

Figure 5Students' opinion on the development of chemistry lessons with the inclusion of experimental activities.

Source: author

Figure 6 shows that students prefer to use experiments to prove theory. Chemistry as an experimental science requires the development of experimental activities or the use of educational resources such as didactic games that validate previously learned theories or content (Paiva, Fonseca, & Colares, 2022).

To the question what is needed to improve the study of chemistry. In this question, values between 1-5 were assigned with different levels of agreement, being: 1 - I totally disagree; 2 - I don't agree; 3 - Undecided; 4 - I agree; 5 - I totally agree.

Table 5Students' responses on how to improve the study of chemistry.

	Relating theory to practice	No need to relate theory to practice	Having a creative teacher	Use of alternative means to explain content

NO.	Valid	47	47	47	47
	Omitted	0	0	0	0
Average		3.57	1.28	3.68	2.91
Median		4.00	1.00	5.00	4.00
Fashion		5	1	5	4
Asymmetry		-1.044	.965	-1.153	-.570
Minimum		0	0	0	0
Maximum		5	4	5	5

As can be seen in table 5, the students' responses to the four variables range from no opinion to totally agree, looking at the minimum and maximum. The average of all the variables is more towards totally agree. However, looking in particular at the variable *"No need to relate theory to practice"*, it can be seen that there is a considerable number of students who totally disagree with the variable. This shows how important it is to relate theory to practice. Therefore, learning content must be associated with skills related to knowing how to do, knowing how to know, knowing how to be and knowing how to be in society (Castro, 2000, cited by Plicas, 2010). Figure 6 shows the percentage distribution related to the variable *"relating theory to practice"* in more detail.

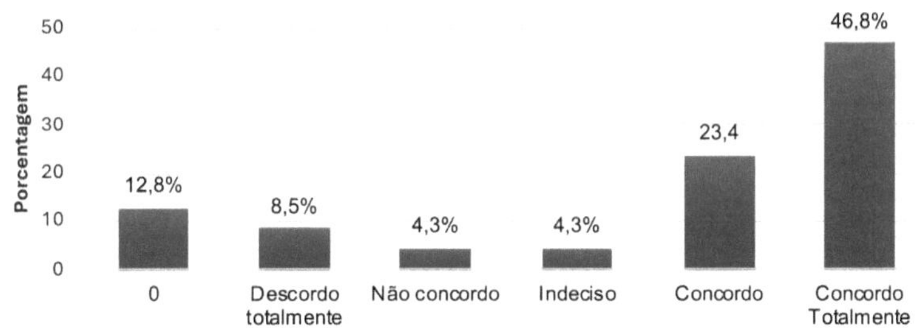

Figure 6Students' opinion on relating theory to practice.

Source: author

On whether teachers carry out experimental activities. Values were assigned, where: 1- Yes- 2-No.

Figure 7Percentage of students' responses to teachers' experimental activities.

Source: author

Figure 8 shows that 83% of the students say they have not had any experimental lessons, which contradicts the answer given by the teachers about carrying out experimental activities in the content taught in the classroom. The lack of experimental activities leads to students' lack of interest in the subject. Therefore, scientific training will only be complete if experimental work is carried out. It is through experimental activities that students benefit from understanding some of the views on the nature of science, which leads to intellectual and conceptual progress and a positive attitude towards science (Lunetta, 1991 as cited in Pacheco, 2015).

The importance of using experimental activities to understand chemistry content. Values were given to reflect the degree of importance, where: 1 - Not important; 2 - Not very important; 3 - Important; 4 - Very important.

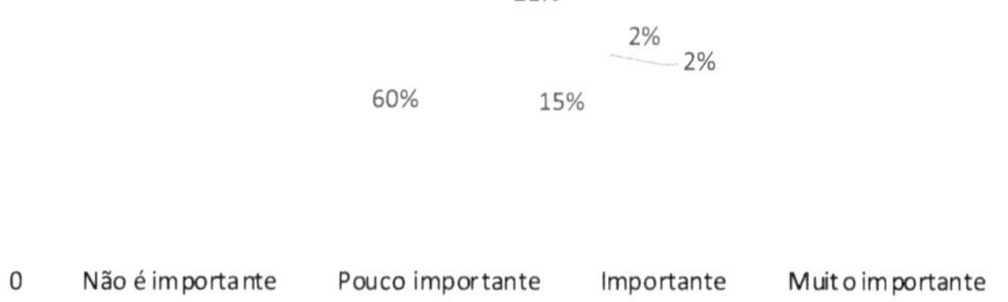

Figure 8Students' opinion on the importance of experimental activities for understanding chemistry content

Source: author

Figure 9 shows that 75% of respondents believe that in order to understand theory, practice is needed to explain phenomena. This corroborates the studies by Marcondes (2006), who considers that the importance of experimental activity in the chemistry learning process is justified if we take into account its pedagogical function of helping students to understand chemical phenomena and concepts. Thus, we agree with Silva (2017, p. 19)who states:

> "By carrying out experimental activities, it is possible to see various benefits in the learning process, among them: the student's active participation in the development of tasks, which will allow them to better assimilate the theoretical content presented in the classroom and awaken the student's interest in identifying scientific processes and phenomena, going through calculations to reach the results."

What experimental activities can provide. In this question, values between 1-5 were assigned with different levels of agreement, being: 1 - I totally disagree; 2 - I don't agree; 3 - Undecided; 4 - I agree; 5 - I totally agree.

Table 6Students' opinion of what experimental activities can provide.

		An environment conducive to more complete learning	More motivation to learn	More instructive teaching	Better understanding of the content	More meaningful learning	Students' handling skills in laboratory work
NO.	Valid	47	47	47	47	47	47
	Omitted	0	0	0	0	0	0
Average		3.45	3.74	3.36	3.64	3.40	3.68
Median		4.00	5.00	4.00	4.00	4.00	5.00
Fashion		5	5	5	5	4	5
Asymmetry		-.999	-1.343	-.901	-1.274	-1.066	-1.262
Minimum		0	0	0	0	0	0
Maximum		5	5	5	5	5	5

Source: author

From Table 6, analyzing the median and mode, all the respondents totally agree with the variables shown, thus demonstrating how important the use of experimental activities is. Experimental activities not only provide the development of scientific knowledge, but also contribute to the development of oral and written skills. The teacher must be an enabler and facilitator of the student's learning so that they take an active role in their own knowledge, in interaction with others and with the environment (Vygostsky, 1998, as cited in Pacheco, 2015, p. 7). It is also important to develop experimental teaching and learning methods, as students will be more active and engaged. Moreira (2012), notes that two conditions are necessary to receive meaningful learning: the learning material is meaningful and that the student has a predisposition to learn.

2.1.3. Analysis of the results of the interview applied to the Chemistry course coordinator.

In order to check the veracity of the answers given by the students and teachers of the 10th grade CFB course, the course coordinator was interviewed. The interview questions are presented below.

As for the characterization of the coordinator, he has a degree in Biology and has been coordinating the Chemistry course for over 7 years. On the question of whether the Joaquim Kapango High School had a laboratory and whether it offered conditions for carrying out experimental activities, he said that there is one that offers working conditions.

The course is made up of 6 teachers who teach General Chemistry in Class 10 in the morning, afternoon and evening. Of the six (6) teachers, only one (1) has a degree in Chemistry.

On the question of whether teachers carry out experimental activities when dealing with the content of the Lê Chatelier principle, the coordinator was clear in replying that no experimental activities have been carried out, although he recognizes the importance of doing so for understanding this content and chemistry in general.

As for the justification, the lack of a guiding document to guide the experimental activity was mentioned. Thus, the research underway will minimize this shortcoming through the proposal drawn up.

Regarding the importance of implementing a didactic strategy for dealing with the Lê Chatelier principle, the coordinator considers it important, as it will help the teaching process as well as the students' learning.

As for the last question on suggestions for improving the performance of experimental activities, the coordinator called for more practice.

2.1.4. Application of the pre-test to the students (sample)

Two sample groups were formed consisting of 47 students each, considered a sufficient sample, since the population is finite (Parra Filho e Santos, 2003; Gil, 1999 as cited in Numbi, 2016, p. 47). A pre-test was therefore administered to the respective sample groups before studying the topic. As a research criterion, the students' answers were scored on a scale of zero to ten (0 to 10), considering that zero (0) does not correspond to an absolute lack of knowledge. The results of the pre-test are reported in Appendix V.

2.2. Didactic strategy for introducing experimental activities in the treatment of content related to the Lê Chatelier principle in the 10th grade of the CFB course at the Joaquim Kapango-Huambo High School

2.2.1. Theoretical foundations of the didactic model

In a teaching context, strategy can be defined as a set of activities that involve selecting, coordinating and preparing the best possible tools and facilitators so that students take responsibility for the knowledge being transmitted. Thus, the term teaching strategies refers to the means used by teachers to describe the learning process according to each activity and expected results.

Paiva et al. (2022)state that the strategies are aimed at analyzing and processing information, solving problems, questioning and reflection, as well as critical thinking, involving two main actors in this process: the teacher as the mediator of knowledge and the student as the person responsible for the action, being the main character, autonomous, participative in the construction of the learning process.

Didactics is one of the branches of pedagogy that focuses on the technique of teaching. Therefore, didactics "studies the teaching technique in all its operational aspects" (Piletti, 2004, p. 42). The term "didactics" arises when adults begin to intervene in the activities of children and young people through the deliberate and planned direction of learning, as opposed to the previous more or less spontaneous forms of intervention (Libânio 2013, as cited in Lima Verde, 2019, p. 52).

Therefore, the implementation of the system-structural-functional method makes it possible to express the logic and sequence of procedures used in the construction of the model, as well as to understand the structure and relationships that constitute the essence of the integration of experimental activity into the dynamics of the chemistry teaching and learning process.

2.3. Teaching model for incorporating experimental activities into the treatment of the Lê Chatelier principle

The model constructed is a simplified theoretical representation of logical connections that simplify the introduction of experimental activities in the teaching-learning process of chemistry, for active, meaningful and solid learning, its construction following the epistemic references of a systematic approach.

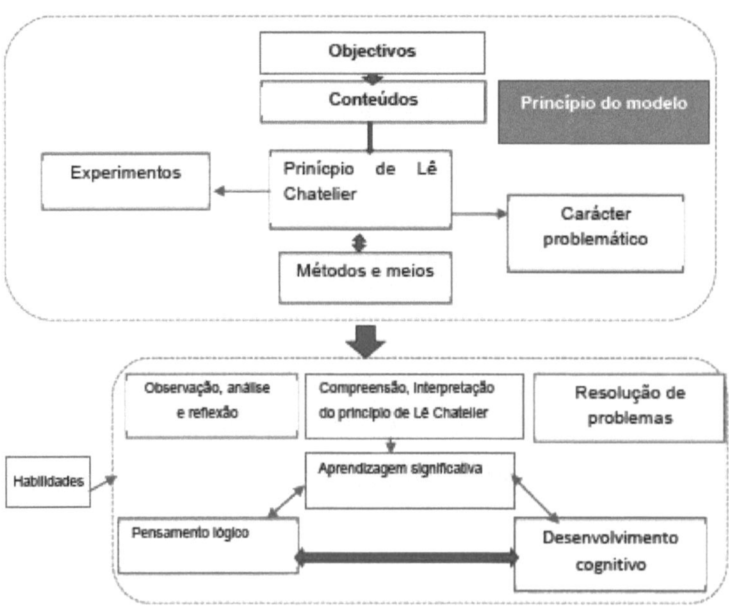

Figure 9Dadactic model, adapted (Sebastião, 2008)

The model starts with the objectives, which represent the guidelines, from which the syllabus emerges, which in this specific case is a question related to the Lê Chatelier principle, which is included in the CBF Grade 10 syllabus.

It takes experiments as its fundamental basis, which allows a close relationship between theory and practice and uses the students' perception of everyday phenomena, allows teachers to manage experiments using materials within their reach and allows students to experiment as well.

The experiments must be planned and directed by the teacher, who must give them a problematic character to allow the students to reflect on the experiments they have observed or carried out.

All the elements mentioned use a set of methods and means, which enable practice to be carried out and indicate "how to do it". This allows the theoretical assumptions of the model to be converted into practical elements.

The model provides for the construction of knowledge on the basis of the formation of cognitive skills, observation, analysis, reflection and others, which enable the resolution of problems related to the Lê Chatelier principle and its applications. The interaction between the elements allows for meaningful learning, the dynamics of which promote the development of logical thinking and cognitive development, as a resulting quality.

2.3.1. Principles of the model

2.3.1.1.   Principle of gradual elevation of the student's self-direction.

The gradual elevation of the student's role in self-directing their own learning is an aspect to consider in the structure and effective development of the teaching-learning process. Some authors argue that the teacher should change their role from providing facts to helping the student discover them for themselves, encouraging them to think for themselves and reach their own conclusions (Garcia, 2000 as cited in Sebastião, 2008, p. 46),

In order to achieve this goal, the solution of the task must support the student in knowing and using different strategies, both for searching for information from different sources and through the observation and interpretation of natural phenomena, as well as in the process of applying them.

In essence, the student's self-direction in the process involves three elements closely related to three components, which are self-diagnosis, self-learning and self-assessment.

Teaching must provide the conditions for students to learn how to learn for themselves, to be able to choose the best paths for their learning; which of them are the most appropriate, given their personal characteristics?

In order to achieve this, students need to be trained in problem-solving to build their knowledge through observation, interpretation and critical analysis, and to use different strategies, both to search for information from different sources and to process and apply it. The selection and systematic use of this strategy will gradually enable the student to acquire greater cognitive independence, in order to form skills, values and attitudes.

Another important aspect of this principle is self-assessment. The teaching-learning process must be conducive to the student being able to self-assess their learning, which is why it is advisable to provide the conditions for them to develop control and evaluation actions through problem solving. When these actions are developed and the student internalizes them, they can then operate a mental plan, being able to anticipate the correct forms of activity, thus achieving a level of self-regulation that allows them to draw up personal projects for self-correction or for making the most of their real potential.

2.3.1.2. Problematic principle of strategy content.

Most researchers agree that every mental process is geared towards solving a problem. The problem determines the objective.

The initiating factor of the mental process is, as a rule, a problematic situation. People start thinking when they feel the need to understand something. Thinking always begins with a problem or a question, with a shadow or a confusion, with a contradiction.

In order to form a critical individual, it is necessary to encourage students to question, giving them not a ready-made explanation of the world, but elements for questioning the various explanations (Galagovsky, et al., 2003 as cited in Sebastião, 2008, p. 46). Therefore, the problem-solving approach aims to precede a change in the school's educational standards, in order to give relevance to the educational act, making it more meaningful for the individual and more intellectually motivating.

The introduction of this type of approach in the classroom will have to take into account what can only be considered real problems that are sufficiently motivating for students to feel committed to solving them.

The didactic models and strategies for the teaching-learning process condition the formation of students' personalities, their value systems, their world view and the way they live. The key to good teaching lies to a large extent in the student's enthusiasm and motivation to learn and the teacher's personal motivation for the act of teaching, in their ability to select methodologies capable of encouraging students to make the intellectual and moral efforts that learning requires on their own initiative, creating in them the "thirst to learn for pleasure and to overcome difficulties".

2.4. Didactic strategy for introducing experimental activities in the treatment of content related to the Lê Chatelier principle

The proposal of a didactic strategy for the treatment of the Lê Chatelier principle through the use of experiments will help to integrate the teaching-learning process in the 2nd Cycle School of the Liceu Joaquim Kapango, that is, in the Physical and Biological Sciences course, helping to consolidate theory and practice, thus enabling them to interpret chemical phenomena and processes quantitatively and qualitatively in the light of this principle, as an important basis for carrying out, understanding and interpreting chemical calculations.

The didactic strategy starts from the objectives, which represent the intention and direction of the teaching-learning process related to the Lê Chatelier principle, from which the syllabus related to this chemical principle derives, which are directly linked to the methods and means that are used to achieve this intention linked to carrying out experiments with a problematic approach.

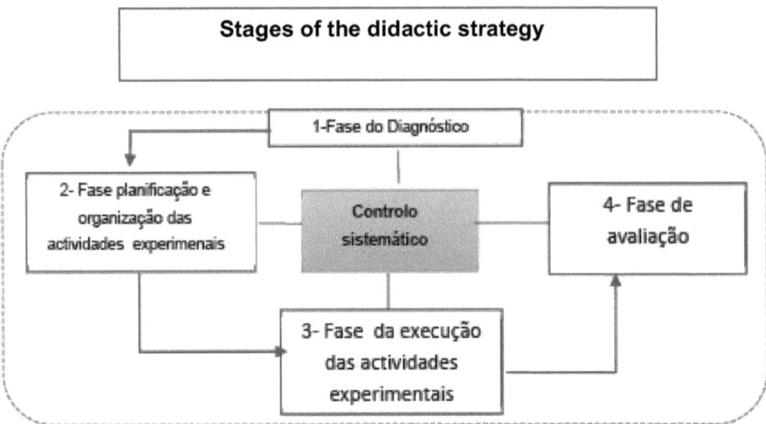

Figure 10Structure of the didactic strategy

## I.	Diagnostic phase

The diagnostic phase made it possible to identify the causes of the shortcomings in relation to the experimental activities, it made it possible to analyze all the variants and needs to ensure activities in the school to improve the teaching-learning process of chemistry, in particular the Lê Chatelier principle.

Another basic element within this phase is the systematic identification and adaptation of the skills to be developed in each topic and each experimental activity to be carried out.

## II.	Phase of the planning and didactic organization of the experimental activities to be developed.

This phase is where the instructive and educational actions that need to be carried out and the dynamic interactions between the subjects involved are specified, namely: teacher-student; highlighting the knowledge that the students are going to learn by selecting and applying different forms of experimental activities in the projects that are planned, carrying the logic of the students' actions in order to learn the required skills.

Its precision is achieved through actions related to the tasks proposed in each activity which, in its system design, should show the progress of the skills acquired by the students

A characteristic of this stage is the precision of the educational potential observed in the process, how much should be achieved in organized planning, which will allow the most appropriate methods to be applied among the possible and required alternatives, which contribute to the development of reflective, critical, creative and innovative thinking, which will generate more meaningful learning for the students.

III.   Execution phase of the planned experimental activities

The essential thing at this stage is to take into account the importance of correct guidance, where the requirements of each activity are actively and flexibly assessed with the students, essentially addressing the output with an emphasis on the formative requirements. The latter must show the degree of responsibility, independence and creativity in decision-making during the development of the different activities.

Due to the complexity of the process of developing experimental activities, the planned activities are divided into three levels, as an expression of the measurement of the students' progress in developing the required skills, delimited by the elements they show: a propaedeutic level, a basic level and an adequate level of performance.

IV.   Phase for assessing the development of skills based on experimentation methods

Assessment based on criteria for developing knowledge, skills and values is the process by which sufficient evidence is gathered of the student's performance in relation to the objectives to be achieved. Its purpose is to determine the level of learning related to the strategy applied.

2.4.1. Implementation of the didactic strategy in the teaching-learning process

Example of experimental activities carried out:

Experiment #1

Lesson time: 45 minutes duration

General objective: To understand the Lê Chatelier principle and the factors that affect a system in chemical equilibrium.

Topic: Chemical equilibrium - Lê Chatelier's principle

Subtheme: Factors affecting chemical equilibrium

Lesson title: Concentration effect

Objective: To demonstrate the effect of the concentration of H^+ on the equilibrium

chromate/dichromate

At the end of this experiment, the student should be able to write an expression for the equilibrium constant; apply Lê Chatelier's principle, identify the ways in which a chemical equilibrium can be affected.

Introduction: Concentration is the ratio between the amount of a solute and the volume of the solution. Many chemical reactions are reversible. In other words, if two chemical species in solution mix and form new species, the new species tend to react to form the original species. The rate of formation of new species will initially be greater than the rate of the opposite reaction. However, with further changes, the rate of formation of new species will equal the rate of the reverse reaction that forms the new substances, thus indicating the top of the equilibrium.

A system can be said to be balanced when it is in chemical equilibrium. A system in equilibrium can be altered or disturbed when the concentration of reactants or products is increased or decreased. In this case, the system tends to shift so that the chemical equilibrium of the reaction is re-established.

Skills: using and caring for instruments; listening, observing, recording, organizing, questioning and explaining the phenomenon that has occurred.

Lesson duration: 45 minutes, **location**: classroom

Previous questions:

1. What factors affect chemical equilibrium?

2. In which direction will the reaction take place if we increase or decrease the concentration of the reactants?

3. In which direction will the reaction take place if we increase or decrease the concentration of the reactants?

Activity: Concentration as one of the factors affecting chemical equilibrium

Type of materials: conventional.

Materials and reagents: $K_2CrO_{4\,(s)}$ (potassium chromate), $K_2CrO_{27(s)}$ (potassium dichromate), NaOH (sodium hydroxide) - 1M concentration, HCl (hydrochloric acid) - 1M concentration, distilled water, 2- 25 ml volumetric flask, spatula, test tube, analytical balance and gloves.

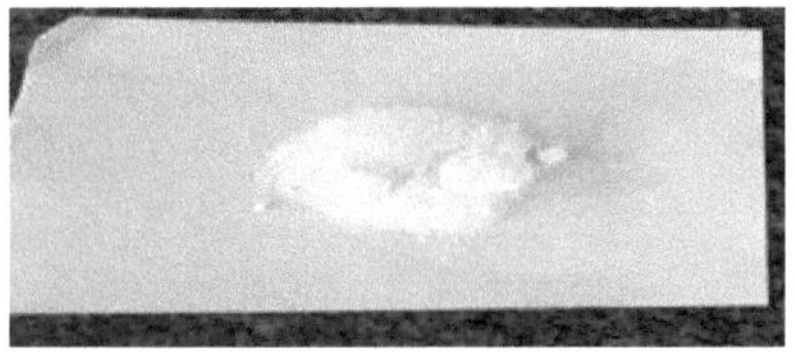

Figure 11 Potassium chromate (K_2CrO_4).

Source: author

Procedures:

1. Measure 2.3530g of potassium chromate and dissolve in 25 ml of distilled water.

Experimental procedure

$$_{aq)} + 2H^{+}{}_{(aq)} \rightleftharpoons CrO72^{-}{}_{(aq)} + H2O_{(l)}$$

At a given temperature, the above reaction is in equilibrium and when the system is disturbed, it is no longer in equilibrium. This is where the principle of Lê Chatelier comes in. What does Lê Chatelier's principle say? Lê Chatelier's principle says that when chemical equilibrium is broken, the system will find a way to establish a new state of equilibrium. These forms of rearrangement are explained by the Lê Chatelier principle.

Example: we can disturb this equilibrium by increasing the concentration of H ions^{+} , the excess of H ions^{+} causes the equilibrium to shift in the right direction, consuming the hydrogen ions and producing dichromate.

In order to compensate for this change, it is necessary to reduce the concentration of H ions$^+$ and this can be done by adding a base. The hydroxyl released by the base consumes the H ions$^+$ and the equilibrium shifts to the left, producing chromate again.

Note: the shift of the equilibrium is detected by the color change.

Explain what happened to the chemical equilibrium after the acid and base were added.

In conclusion, the given reaction obeys the Lê Chatelier principle, which establishes concentration as one of the factors influencing the equilibrium of a system.

Experiment #2

Topic: Chemical equilibrium - Lê Chatelier's principle
Subtheme: Factors affecting chemical equilibrium
Lesson title: effect of pressure

Objective: To demonstrate in practice the influence of pressure on chemical reactions.

> At the end of this experiment, the student should be able to write an expression for the equilibrium constant; apply Lê Chatelier's principle, identify the ways in which a chemical equilibrium can be affected.

Introduction: Pressure (p) is the ratio between force and its distributed area. The term pressure is used in various branches of science as a scalar quantity that measures the action of one or more forces on a given space, which can be liquid, gaseous or even solid. Pressure is an inherent property of any system and can be beneficial or unfavorable to human beings.

Skills: using and caring for instruments; listening, observing, recording, organizing, questioning and explaining the phenomenon that has occurred.

Lesson duration: 45 minutes, **location:** laboratory

Previous questions

1. What happens with the increase and decrease of pressure in a reaction?
2. What happens to the increase and decrease of volume in a reaction?

Activity: pressure as one of the factors affecting chemical equilibrium

Summary: effect of pressure

Type of materials: low cost

Reagents and materials: 2 complete syringes, 50 ml precipitation beaker, nail, match, soda water and methyl red or orange indicator.

Experimental procedure:

$$H_2O_{(g)} + C_{14}H_{14}N_3NaO_2S_{(aq)} \longleftrightarrow C_{14}H_{13}N_3NaO_2S^{\square}_{(aq)} + H_3O^+_{(aq)}$$

a) Pour the soda water into the precipitation beaker, then add 3 drops of the indicator, an acidic colored solution will form (image below).

Figure 12Soda water + methyl orange indicator.

Source: author

b) Use syringes A and B to draw 1 or 2 ml of the solution. Then cap the syringes.

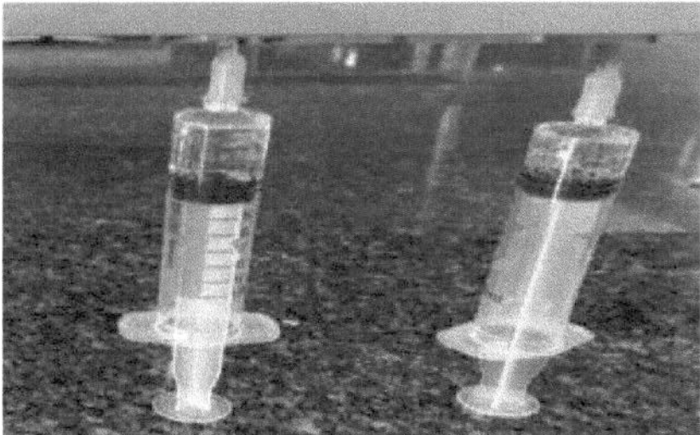

Figure 13Demonstration of the effect of pressure A and B.

A B

c) Pull the plunger of syringe B all the way out, then hold it. Wait until the bubbles in syringe B disappear.

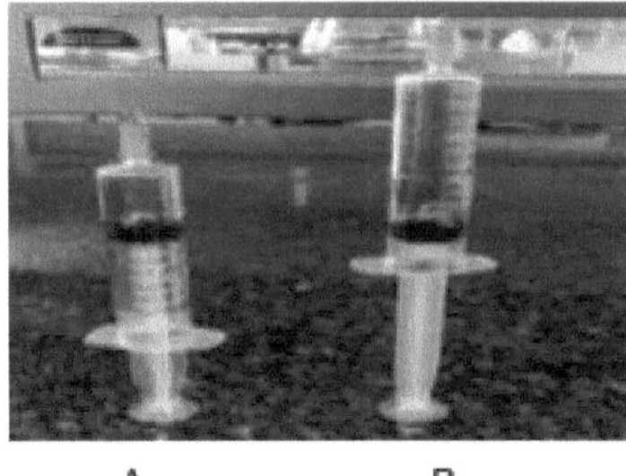

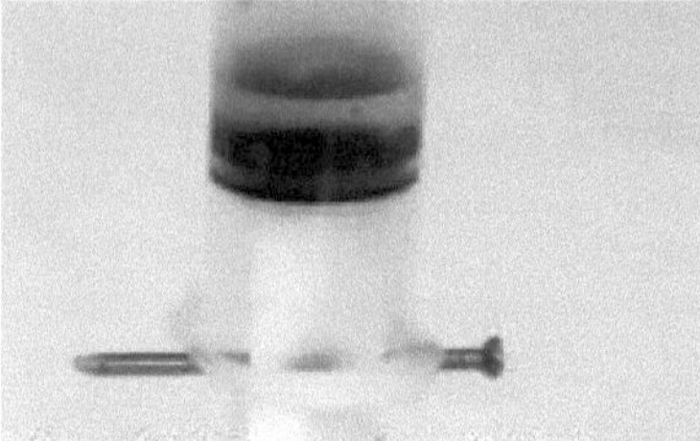

Figura 13.1.1. Como prender o êmbolo (aquecer o metal e perfurar o êmbolo antes de colocar a solução).

A   B

Figura 13.1. Demonstração de efeito da pressão.

d) Remove the cap from syringe B and lower the plunger to the initial stage and cap again.

AB

Figure 13.2: Demonstration of the effect of pressure.

e) Compare the solution in syringe A and B.

When comparing syringe A and B, it can be seen that solution B appears darker, and this is due to the decrease in pressure, making it less acidic than A.

In conclusion, the variation of the solution from acidic to less acidic is visible, which is due to the decrease in pressure. In other words, pressure influences the equilibrium of a system. Therefore, it can be said that variations in pressure also alter the equilibrium constants.

Experiment #3

Topic: Chemical equilibrium - Lê Chatelier's principle

Sub-theme: Factors affecting chemical equilibrium

Lesson title: Effect of temperature

Objective: To demonstrate in practice the influence of temperature on chemical reactions.

> At the end of this experiment, the student should be able to write an expression for the equilibrium constant; apply Lê Chatelier's principle, identify the ways in which a chemical equilibrium can be affected.

Introduction: Temperature is the average kinetic energy of each of the particles in a system. In particular cases, temperature is defined only for systems in thermal equilibrium. The International System of Units (SI) establishes a special scale for absolute temperature, which is 273.16 K. Where K is Kelvin.

Skills: using and caring for instruments; listening, observing, recording, organizing, questioning and explaining the phenomenon that has occurred.

Lesson duration: 45 minutes, **location**: classroom

Previous questions

1. What happens as the temperature rises and falls in a reaction?
2. What state of aggregation will the substances have as the temperature rises and falls in a reaction?

2. Activity: Concentration as one of the factors affecting chemical equilibrium

Summary: effect of temperature

The system in equilibrium can be altered or disturbed when the temperature in the system increases or decreases.

When energy is added to or removed from a chemical system, the system resists the change by absorbing or releasing energy so that equilibrium is restored.

When the temperature of the system changes, the chemical equilibrium changes as follows:

An increase in temperature promotes an endothermic reaction and the system absorbs heat.

When the temperature is reduced, an exothermic reaction is favored and the system releases heat:

$$\left[Co(H_2O)_6\right]^{+2}_{(aq)} + 4Cl^-_{(aq)} \rightleftharpoons \left[CoCl_4\right]^{2-}_{(aq)} + 6H_2O_{(l)}$$

Cor de rosa

Iões hexahidrato de cobalto II

Cor azul

Iões tetracloreto de cobalto II

Type of materials: Conventional

Reagents and materials: $CoCl_2$ (cobalt chloride), HCl (hydrochloric acid) - 1M concentration, distilled water, 50 ml volumetric flask, spatula, 3 test tubes, analytical balance or other and gloves, 100 ml precipitation beaker.

Procedures:

    a) Measure out 2.75g of cobalt chloride;

    b) Pour the measured cobalt chloride into the 50 ml volumetric flask ;

    c) Add a few drops of distilled water to the volumetric flask, then shake.

    d) When the cobalt chloride has dissolved, add more water up to the limit line of the volumetric flask and shake again;

    e) Transfer 4 to 8 ml of the prepared solution into a test tube;

    f) Add drops of HCl to the prepared solution and stir until it changes color (violet);

How can a change in temperature affect the balance?

1. Transfer 2 to 4 ml of the prepared solution into test tubes A and B.
2. Prepare 2 precipitation beakers A and B of 100 ml, in A put hot water (high temperature), in B put cold water (low temperature).
3. Place test tube A in precipitation beaker A and test tube B in precipitation beaker B. Observe what happens.

Note: when test tube A is placed in the precipitation beaker containing hot water, the temperature of the system rises and the equilibrium shifts towards the formation of more products.

This phenomenon occurs because the direct reaction is endothermic and the system will be restored by absorbing heat.

In conclusion, the given reaction obeys Lê Chatalier's principle, which establishes temperature as one of the factors that influence or affect the equilibrium of a system. Furthermore, it can be said that variations in temperature also alter the equilibrium constants.

Methodological guidelines

The systems being studied in this subdomain must be homogeneous, gaseous or aqueous. With regard to the quantitative aspects of chemical equilibrium, cases in which only data relating to the initial composition of the system is provided should be excluded.

The use of experimental activities will help students understand that a chemical system can have an infinite number of equilibrium states at the same temperature with the same equilibrium constant. In addition, they can also help to understand the evolution of chemical systems as a result of disturbances in chemical equilibrium, with the advantage of being able to examine macroscopically what happens in these cases, reinforcing the idea of the dynamic nature of chemical equilibrium.

Questionnaire

What factors affect chemical equilibrium?
How does pressure affect chemical equilibrium?
How does temperature affect chemical equilibrium?
When should we apply the Lê Chatelier principle?

In the **fourth stage**, the teacher asked the students to summarize the activities carried out and at the end present a report on the activities carried out. To get a clear idea of whether there had been any progress in learning, the teacher gave the students a post-test, which served as the **fifth stage**. Other suggested activities (Appendix VII).

2.5. Application of post-tests to the students involved (experimental group)

After the intervention in the working groups, the post-test was administered only to the students who had taken the pre-test in the specified classes. These tests were administered by the author of this paper in the presence of the class teachers.

Likewise, at the end of the intervention in the experimental class, the class teacher was interviewed about the work carried out. The interview yielded the following result: the strategy had a positive impact on the student's learning, as it is based on a set of procedures which, when rigorously observed, lead the student to what is expected, and when applied correctly, contribute to a good understanding of the concepts, as well as to problem-solving skills, so we advise teachers to try to exploit the benefits of this strategy as much as possible in order to minimize or overcome difficulties in problem-solving. The author of this paper agrees with these statements, as he was present and observed the level of participation and enthusiasm of the students. The results of the post-test (Appendix VI) are shown in more detail in Chapter III.

Conclusion of chapter II

The incorporation of experimental activities as didactic strategy enhances the process of teaching chemistry, as this practice provides many benefits that allow students to understand chemical phenomena.

The proposed didactic strategy is a methodological alternative for the chemistry teaching-learning process, with experimental activities that promote reflection, the exchange of ideas and the development of active and meaningful learning.

CHAPTER III. VALIDATION OF THE TEACHING STRATEGY USING THE COMPARATIVE MANN-WHITNEY METHOD

Chapter III. Validation of the teaching strategy using the Mann- Whitney comparative method

In order to obtain evaluation criteria regarding the didactic aspects determined and the feasibility of applying the proposed didactic strategy, the comparative method was used, based on the results of the post-tests applied to the samples (control and experimental groups).

The aim of the comparative method is to use comparison - through similarities or differences - to gain new knowledge. According to Coelho (2022)in general, the comparative method involves establishing parallels between two or more objects of study in order to analyze similarities and differences. The author adds that it is a method of proving or disproving theories and hypotheses based on comparisons.

3.1. Results of the student questionnaire after applying the didactic strategy

In order to improve the students' learning during the experimental activities, taking into account the content presented, demonstrated and discussed, a questionnaire was drawn up and answered by 23 students representing the experimental group. The results of the students' responses are shown in Figures 14 and 15.

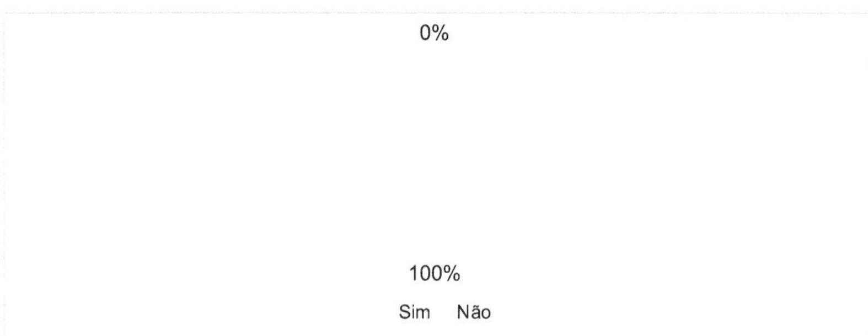

Figure 14: Students' opinion of their enjoyment of taking part in lessons with experimental activities.

Figure 14 shows that all 23 students, which corresponds to (100%), enjoyed the activities carried out, which led to greater application of the content studied and greater attention in class.

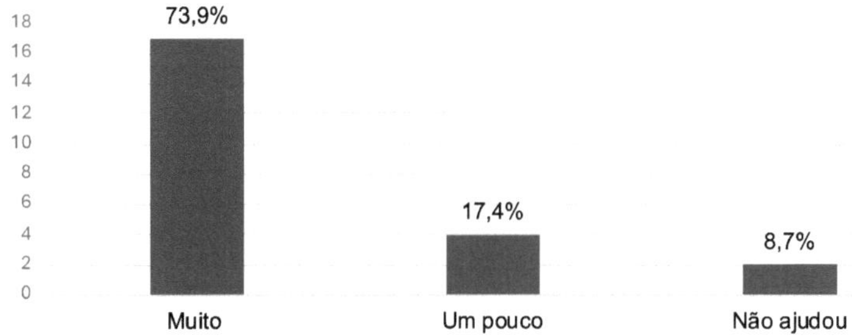

Figure 15: Students' opinion of experimental activities' help in understanding the factors that affect a system's equilibrium.

Figure 15 shows that 73.9% of the students felt that carrying out the experimental activities was very helpful in understanding the topics, indicating that there was an understanding of the application of Lê Chatelier's principle, how temperature, concentration and pressure affect the equilibrium of a system, as well as the importance of this principle in biological, industrial and other processes.

3.2. Validation of the didactic strategy for introducing experimental activities in the treatment of the Lê Chatelier principle

In order to compare equivalence between the groups, the normality test was initially determined to check the distribution of the sample data. Details are shown in Table 7.

Table 7Results of the normality test.

	Kolmogorov-Smirnova			Shapiro-Wilk		
	Statistics	df	Sig.	Statistics	df	Sig.
What happens as the temperature rises and falls in a reaction?	.501	18	.000	.457	18	.000
Effect of concentrations:	.523	18	.000	.373	18	.000
What is the catalytic converter for?	.392	18	.000	.624	18	.000
When should the Lê Chatelier principle be applied?	.463	18	.000	.552	18	.000
Lilliefors Correlation of Significance						

Table 7 shows that the significance level (sig) is less than 0.05, so the data is non-parametric. Therefore, to compare the averages between the groups, the mean rank was analyzed as shown in table 8.

Table 8group rank results

Posts				
	Class	N	Medium post	Sum of Ratings
What happens as the temperature rises and falls in a reaction?	Control group	18	15.25	274.50
	Group exp.	23	25.50	586.50
Effect of concentrations	Control group	18	13.22	238.00
	Group exp.	23	27.09	623.00
What is the catalytic converter for?	Control group	18	20.83	375.00
	Group exp.	23	21.13	486.00
When should the Lê Chatelier principle be applied?	Control group	18	13.53	243.50
	Group exp.	23	26.85	617.50
	Total	41		

Table 8 shows the average rank and sum of ranks between the groups, where the experimental group has the highest rank and the highest average. The Mann-Whitney test was used to determine whether there was a statistical difference between the groups. Details are shown in table 9.

Table 9Statistical results of the Mann-Whitney test

Test statistics[a]				
	What happens as the temperature rises and falls in a reaction?	Effect of concentrations:	What is the catalytic converter for?	When should the Lê Chatelier principle be applied?
Mann-Whitney U	103.500	67.000	204.000	72.500
Wilcoxon W	274.500	238.000	375.000	243.500
Z	-3.046	-4.179	-.093	-3.960
Significance Sig. (bilateral)	.002	.000	.926	.000
a Grouping Variable: Class				

In table 9, the significance level (sig) for the Mann-Whitney test shows that there is a statistically significant difference between the control and experimental groups ((U = 103.500; p < 0.002); (U = 67.000; p < 0.000); (U = 72.500; p < 0.000)), with the exception of the variable "what the catalyst is for" where (U = 204.500; p > 0.926), which reflects the equality in the group averages. Figures 17, 18, 19 and 20 show more details on the results of the groups' U-test.

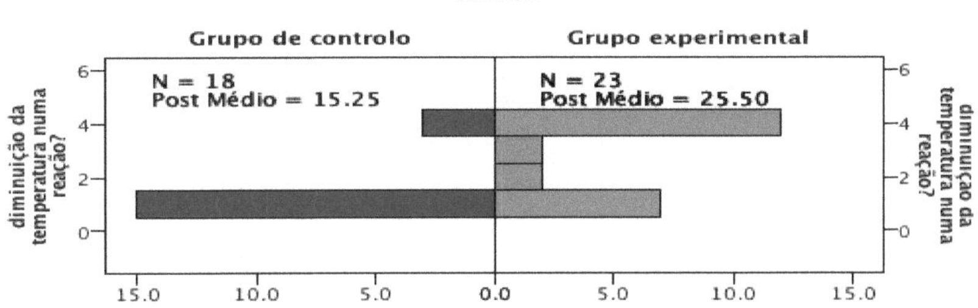

Figure 16Result of the students' response to what happens when the temperature increases and decreases in a reaction (U = 103.500; p < 0.002).

Source: author

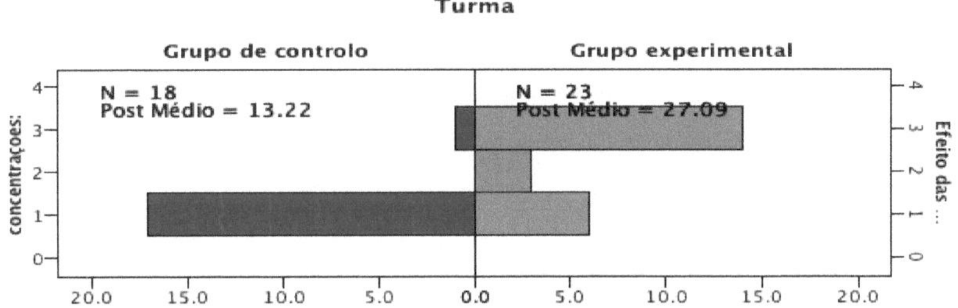

Figure 17Result of the students' response to the effect of concentrations (U = 67.000; p < 0.000).

Source: author

Figure 18Result of the students' response on what the catalyst is used for (U = 204.500; p > 0.926).

Source: author

Figure 19Result of the students' response on *when the Lê Chatelier principle should be applied* (U = 72.500; p < 0.000).

Source: author

When comparing the results of the post-test averages between the groups, a statistically significant difference was noted, with the experimental group (new method) having higher results than the control group (traditional method). Corroborating Barbosa (2016)(2016), learning in the school environment, the choice of teaching resources and teaching strategies influence the construction of knowledge, aiming at school educational practice, especially when seeking to exclude or minimize students' difficulties in understanding the concepts exposed. In this way, it can be said that the didactic strategy for introducing experimental activities in the treatment of the content of the Lê Chatelier principle contributes significantly to the teaching-learning process of chemistry, particularly in the learning of the principle.

Conclusion of chapter III

When comparing the post-test results between the groups using the Mann-Whitney comparison test, a statistically significant difference was found. There was a significant improvement in the PEA of the Lê Chatelier principle, where the experimental group (new method) had higher results than the control group (traditional method).

CONCLUSIONS AND RECOMMENDATIONS

IV. Conclusions

From the diagnoses made, it emerged that the difficulties faced by students in the 10th grade of the CFB course in understanding the content of Chemistry, in particular the subject under study, can be overcome or minimized by carrying out experimental activities.

Both teachers and students are victims of this lack of experimental activities.

The didactic proposal under discussion is a didactic alternative to the chemistry teaching-learning process, particularly the principle de Lê Chatelier which integrates experimental activities to enhance students' active and meaningful learning.

The teaching strategy was validated using the Mann-Whitney comparative test, which found a statistically significant difference between the groups. There was a significant improvement in the PEA of the Lê Chatelier principle, where the experimental group (new method) had better results than the control group (traditional method).

.

V.   Recommendations

Given the relevance of the subject, we suggest that other researchers continue the study, using other methods such as computer simulations, as a prerequisite for a contextualized and integrated teaching-learning process to minimize students' learning shortcomings. They could also expand and diversify the study sample, the study topic and increase the study time.

We recommend using the didactic strategy presented in this work as a basis and adjusting it according to the context in order to deepen the students' skills by including experimental activities in the treatment of the Lê Chatelier principle for 10th graders in the CFB course at the Joaquim Kapango-Huambo High School.

BIBLIOGRAPHICAL REFERENCES

VI.   Bibliographical references

Araújo, M. S., & Abib, M. L. (2003). *Experimental Activities in Physics Teaching: different approaches, different purposes. Revista Brasileira de Ensino de Física, v.25, n.2.*

Barbosa , A. C. (2016). *Reading mediation of didactic texts in Chemistry classes: An approach focused on the Enem reference matrix. Ensaio Pesquisa em Educação em Ciências.* Belo Horizonte.

Bauman, N. E. (2015). Application of Le Chatelier-Brown principle to interpret the results of long-term measurements of voltage fluctuations in small volumes of electrolyte Glagolev K.V.1 , Morozov A.N.1. *Electron. journal 2015. ., 6*, P. 1-9.

Bedin , E., & Cassol , C. (2016). *Teaching Chemical Equilibrium in Basic Education: from analogies to experimental activities,.* XVIII ENEQ, .

Behrens, M. A. (2005). *The emerging paradigm and pedagogical practice. Petrópolis: Vozes.*

Butts, B., & Smith, R. (1987). *What do students perceive as difficult in H.S.C. Chemistry? The Australian Science Teachers' Journal.*

Canzian, R., & Maximiano, F. A. (2010). *Le Chatelier Principle: What Has Been Presented in Textbooks?* Brazil.

Canzian, R., & Maximiano, F. A. (2009). *Analysis of the Formulation of Le Chatelier's Principle in Textbooks.* Corritiba.

Carvalho, A. M. (2005). *Science in Primary School: physical knowledge. Scipione.199p.* São Paulo.

Catumbela , F. C. (2016). *learning concepts through demonstrative experiments.* Dissertation, ISCED-HUILA, Exact Sciences, Lubango.

Coelho, B. (December 30, 2022). *mettzer.* Retrieved August 27, 2023, from Comparative method: create comparisons to understand what things are (and aren't!): https://blog.mettzer.com/metodo-comparativo/

Farias , e. (2017). *Study of chemical equilibrium .* SP.

Ferreira, L. H., Romeu , C. R.-F., & Dácio, H. H. (1997). *Some Simple Experiments Involving Le Chatelier's Principle.*

Figuerêdo, A. M. (2018) . *Contextualized Experiment on Chemical Equilibrium for High School Class. INTERNATIONAL JOURNAL EDUCATION AND TEACHING (PDVL) ISSN 2595-2498*.

Finley, F. N., Stewart, J., & Yarroch, W. L. (1982). *Teachers' perceptions of important and difficult science content. Science Education*.

Freire, M. d., Júnior, G. A., & Silva, M. G. (2011). *Overview of problem solving and its applications in chemistry teaching. Acta scientiae, v.13, n.1*.

Gonçalves, F. P. (2005). *The experimentation text in chemistry education: pedagogical and epistemological discourse.* Federal University of Santa Catarina, Brazil.

Guimarães , C. C. (2009). *Experimentation in chemistry teaching: Paths and detours towards meaningful learning. Química Nova na Escola*.

Hodson , D. (1994). *Hacia un enfoque más critico del trabajo de laboratorio. Enseñanza de las ciencias*.

INIDE (2013). *Curriculum for the 2nd cycle of general secondary education.* Angola: Editora Moderna. Educational reform. 3rd Edition.

INIDE (2014). *10th, 11th and 12th Grade Program. IIº Ciclo do Ensino Secundário Geral.* Luanda.

INIDE, & MED. (2020). *Minimum Chemistry Program 10th, 11th and 12th Classes. 2nd Cycle of General Secondary Education.* Luanda-Angola: Pre-press, Printing and Finishing.

João, J. D. (2007). *Methodological proposal for the introduction of problematic demonstrative experiments in the teaching of physics at Lubango middle school.* ISCED-Huíla, Lubango.

Junior, J. G. (2017). *The approach to the topic of chemical equilibrium in the lesson plans of future teachers.* Sevilla.

Law no. 32/20 of August 12 amending law no. 17/16, d. 7. (2020). *Bases of the education and teaching system.* Angola: Official Gazette.

Libaneo, J. C. (2006). *Didactica.* São Paulo, Brazil: Cortez.

Lima Verde , E. S. (2019). *Didactics and its object of study.* Federal University of PIAUI, Brazil.

Lima, J. O. (2012). Perspectives of new technologies in the teaching of chemistry. *Revista espaço académico*.

Marcondes, M. (2006). *Thematic workshops in public education aimed at continuing teacher training. GEPEQ - Research Group in Chemical Education of the Chemistry Institute of the University of São Paulo.* São Paulo: GEPEQ - Grupo de Pesquisa em Educação Química do Instituto de Química da Universidade deSão Paulo.

Mol, G. S. (2017). *Qualitative Research in Chemistry Teaching, Revista Pesquisa Qualitativa.*

Moreira , M. A. (1986). *The textbook as a vehicle for curricular emphases.* UFRGS Physics Institute, Porto Alegre, Brazil.

Mouras, J. F., da Silva, H. S., & Teixeira, J. G. (2010). *study with high school students and future chemistry teachers on the application of Le Chatelier's principle.* Universidade Federal de Uberlandia, aqueline Fernandes Mouras, Heliene Sousa da Silva, José Gonçalves Teixeira. Brasília: EAP.

Mouras, J. F., da Silva, H. S., & Teixeira, J. G. (2010). *study with high school students and future chemistry teachers on the application of Le Chatelier's principle.* Federal University of Uberlandia. Brasília: EAP.

Nerci, I. G. (1989). *Teaching Methodology, An Introduction. 3rd Edition. .* São Paulo: Editora Atlas.

Numbi , F. T. (2016). *Teaching chemical reactions through problem solving in experimental activities.* Dissertation, ISCED-Huíla, Lubango.

Ogasawara, J. S. (2009). *Skinner and Vygotsky's concept of learning: a possible dialog.* Monograph, State University of Bahia, Salvador.

Oliveira, J. R. (2010). *Contributions and approaches to experimental activities in science teaching: gathering elements for teaching practice. In: Acta Scientiae Canoas v. 12 n.1 .*

Paiva, M. M., Fonseca, A. M., & Colares, R. P. (2022). Potentiating didactic strategies in the teaching-learning of chemistry. *Revista de Estudos em Educação e Diversidade, 3*, p. 1-25.

Paiva, M. M., Fonseca, A. M., & Colares, R. P. (Jan./Mar. 2022). Potentiating didactic strategies in chemistry teaching and learning. *Revista de Estudos em Educação e Diversidade, 3*, p. 1-25.

Piletti, C. (2004). *Didáctica Geral* (23 ed.). Campinas, São Paulo, Brazil: Ética.

Queliz, J. (1995-1998). A formulation for a principle: historical analysis of Le Chatelier's principle. *Revista mexicana de fisica*.

Quílez, P. J., Solaz, P. J., Castelló, H. M., & Sanjo, S. L. (1993). *La necesidad de un cambio metodológico en la enseñanza del equilibrio químico: limitaciones del principio de le chatelier. Science Teaching*.

Quílez-Pardo, J. (March 9, 1995). Una formulación para un principio: análisis histórico del principio de Le Chatelier. *Revista Mexicana de Física*.

Rodrigues, E. (2015). *Teaching and learning process in middle schools in Angola*. São Paulo: Química Nova.

Saad, F. D. (2005). *Demonstrations in Science: exploring air and liquid pressure phenomena through simple experiments*. São Paulo: Livraria da Física.

Sales, D. M., & Silva, F. P. (2010). *Using experimental activities as a science teaching strategy. Senac Faculty Teaching, Research and Extension Meeting*.

Sebastião, A. (2008). *Methodological strategy for introducing an experimental activity in the treatment of the law of conservation of mass*. Huíla: ISCED-Huíla.

Silva, D. (2021). *International, Chemical equilibrium: teaching and learning trends in scientific publications based on an analysis of national and international journals*. Ouro Preto: Ouro Preto State University.

Silva, E. D. (2017). *The importance of experimental activities in education*. Candido Mendes University-Integrated College. Rio de Janeiro: AVM.

Sousa, D. (2015). *Pedagogical practice in chemistry. :*. Belém: Editaedi.

Tomalela, M. (1998). *Concept Maps and Experimental Work as Meaningful Learning Activities in Chemistry: A Study in the 10th Grade of the Normal Institute of Education of Lubango*. PhD thesis, University of Minho, Portugal.

Wongombo, H. H. (2015). *Methodological strategy for the development of experimental activities in the teaching-learning process of chemistry. Study carried out in the secondary schools of "Lubango", "nambambi" and "arimba"*. Dissertation, Huila.

Zanon, L. B., & Silva, L. H. (2000). *Experimentation in science teaching. In: Schnetzer, R. P.; Aragão, R. M. R. (Orgs.) Ensino de Ciências: fundamentos e abordagens. Campinas: V Gráfica.*

APPENDICES AND ANNEXES

APPENDIX I

QUESTIONNAIRE SURVEYS FOR STUDENTS

This questionnaire survey is aimed at Grade 10 students in the Physical and Biological Sciences course at Liceo Joaquim Kapango-Huambo. Its aim is to collect information on the use of experimental activities, particularly in the treatment of the content "Lê Chatelier's principle" in this school, which in turn will serve as the basis for the preparation of the dissertation with the theme: didactic strategy for the introduction of experimental activities in the treatment of Lê Chatelier's Principle.

The survey is anonymous, so you are expected to answer the questions below truthfully:

I.   Characterization of the student:

     a) Personal details: Mark your answer with an X

     b) Sex: male _______; female ______Age: _________anos.

II.   Chemistry PEA

1.   Do you like chemistry?

I don't like it ( ) a little ( ) I like it ( ). Justify your statement

____

2.   How would you like chemistry lessons to be developed? (Mark with an X indicating a value between 1-5 in each point of the table below, where: 1 - Totally disagree; 2- Don't agree; 3 Undecided; 4 - Agree; 5 - Totally agree).

NO.	Development	1	2	3	4	5
A	Based on exercises (use of formulas)					
B	No experimental activities included					
C	With the inclusion of experimental activities					

3.   Do you think that in order to improve the study of Chemistry it is necessary to: (Mark with an X indicating a value between 1-5 in each point of the table below, where: 1 - Totally disagree; 2- Don't agree; 3 Undecided; 4 - Agree; 5 - Totally agree).

72

N O.	Required:	1	2	3	4	5
A	Relating theory to practice					
B	No need to relate theory to practice					
C	Having a creative teacher					
D	Use of alternative means to explain content					

III. Perception of experimental activities

4. Have teachers carried out experimental activities in chemistry lessons? YES ( ); NO ( ). If you answer yes, answer questions 6, 7 and 8. If you answer no, go on to the next questions.

5. How often per term have you carried out experimental activities in class? Once ( ), twice ( ), three ( ) four ( ), other ( ).

6. Were the lessons involving experimental activities?

Poor ( ); bad ( ); normal ( ); good ( ); excellent ( ) . Justifique a sua afirmação--

7. Where did the experimental activities take place?

In the classroom ( ); in the laboratory ( ); in the schoolyard ( ); elsewhere (

8. In your opinion, is it important to carry out experimental activities in order to understand chemistry content?

not important (); not very important (); important (); very important (

9. Do you agree that experimental activities can provide: (Mark with an X indicating a value between 1-5 in each point of the table below, where: 1 - Strongly disagree; 2- Do not agree; 3 Undecided; 4 - Agree; 5 - Strongly agree).

N O.	Experimental activities	1	2	3	4	5
A	An environment conducive to more complete learning					
B	More motivation to learn					
C	More instructive teaching					
D	Better understanding of the content					
E	More meaningful learning					
F	Students' handling skills in laboratory work					

Thank you very much for your cooperation!

The researcher: Manuel Pina

QUESTIONNAIRE SURVEYS FOR TEACHERS

This survey is aimed at Grade 10 Chemistry teachers in the Physical and Biological Sciences course at Liceo Joaquim Kapango-Huambo. It aims to collect information on the use of experimental activities in the treatment of the "Lê Chatelier principle" content at this school. The data collected will support the preparation of the dissertation entitled: didactic strategy for the introduction of experimental activities in the treatment of Lê Chatelier's principle.

The survey is anonymous, so you are expected to answer the questions below truthfully:

I. **Characterization of the teacher:**

c) Personal details: Mark your answer with an X

d) Sex: male ______; female ______Age: _________anos.

e) Academic qualifications: Secondary technician:___;Bachelor___;Licentiate___;Master___; Doctor___.

f) Training area: _______________________________

g) How long have you worked as a chemistry teacher? _____ years.

II. **Characterization of the chemistry teaching-learning process**

▪ In general, how would you characterize the current state of the chemistry EAP?

____

____

III. **Perception of experimental activities and the Lê Chatelier principle**

1. Dear teacher, in general, have you developed experimental activities? Yes ( ) no ( ).

2. Have you carried out experimental activities when dealing with the "Lê Chatelier Principle" content?   Yes ( ) no ( ), If you answered yes to questions 1 and 2, answer questions 3 and 4, otherwise go on to the next questions.

3. How often per quarter do you carry out experimental activities?

Once ( ), twice ( ), three ( ) four ( ), five ( ), more ( ).

4. What types of experimental activities do you use in the Chemistry PEA? Demonstrative ( ); verifying ( ); investigative ( ); other ( ).

5. What difficulties have you encountered in carrying out experimental activities in the treatment of the "Lê Chatelier principle" content? (Mark with an X indicating a value between 1-5 in each point of the table below, where: 1 - I totally disagree; 2- I don't agree; 3 Undecided; 4 - I agree; 5 - I totally agree).

NO.	Difficulties	1	2	3	4	5
A	Lack of materials and reagents					
B	Lack of a document to guide experimental activities					
C	Associate experiment with content					
D	Selection of materials and reagents					

6. In your opinion, does the experimental activity contribute to the construction of students' knowledge? Rarely ( ); always ( ); never ( ). Please justify your answer:____

—

7. In your opinion, is it important to carry out experimental activities in order to understand the "Lê Chatelier Principle" content?

Not important ( ); important ( ); very important ( ).

8. Is there encouragement from the school management to carry out experiments in chemistry lessons? Rarely ( ); always ( ); never ( ).

9. In the last 3 semesters, have you attended a training workshop on conducting experimental activity? Yes ( ) no ( )

10. To what extent do you consider it important to implement a didactic strategy for introducing experimental activities in the treatment of the "Lê chatelier principle" content?

Not important ( ); not very important ( ); important ( ); very important ().
Please justify your choice:

____

11. According to your teaching experience in chemistry, do you think that experimental activities can provide: (Mark with an X indicating a value between

1-5 in each point of the table below, where: 1 - I totally disagree; 2- I don't agree;
3 Undecided; 4 - I agree; 5 - I totally agree).

N O.	experimental activities	1	2	3	4	5
A	An environment conducive to more complete learning					
B	More motivation to learn					
C	More instructive teaching					
D	Better understanding of the content					
E	More meaningful learning					
F	Students' handling skills in laboratory work					

Recommendations   **or**   **suggestions:**

____

____

Thank you very much for your cooperation.

The Researcher: Manuel Pina

APPENDIX III

INTERVIEW WITH THE CHEMISTRY COORDINATOR

I.   **Characterization of the coordinator:**

    h) Personal details: Mark your answer with an X

    i) Sex: male _______; female ______Age: _________anos.

    j) Academic qualifications: Secondary technician:___;Bachelor___;Licentiate___;Master___; Doctor ___.

    k) What is your area of training?

       ____

    l) How long have you worked as a coordinator? _____ years.

II.   **Characterization of the Lyceum**

A. Does the school have a laboratory? YES ( ) NO ( ), if YES, does it offer working conditions? YES ( ) NO ( ).

B. Is there a shortage of reagents? YES ( ) NO ( ).

C. How many teachers teach Grade 10 at the CFB?_______ How many have a degree in Chemistry?____________ and what are the periods? ____________________________.

D. In general, how do you rate the Grade 10 Chemistry program at CFB?____

____

III.   **Characterization of Experimental Activities.**

1. Have teachers carried out experimental activities in general? YES ( ) NO ( ), If YES, how often per term?

2. Have teachers carried out experimental activities when dealing with the Lê Chatelier principle? YES ( ) NO ( ).

3. If YES, do you have a guide or guidebook for this? YES ( ) NO ( ). If NO to question 3, please justify your statement

____

____

4. In the last 3 semesters, have teachers at your institution, particularly chemistry teachers, taken part in training seminars related to experimental activities? YES ( ) NO ( ).

78

5. What difficulties have you encountered in carrying out experimental activities on the "Lê Chatelier principle"?

6. Do you consider it important to implement a didactic strategy for the development of experimental activities in the treatment of the Lê Chatelier principle? YES ( ) NO( ). Justify your statement

____

____

7. What difficulties does the institution face in implementing experimental activities in general?

____

____

8. What suggestions do you have for improving experimental performance?

____

____

Thank you very much for your cooperation!

The Researcher: Manuel Pina

APPENDIX IV

PRE-TEST AND POST-TEST

Age_______ years; Sex M_____/F_____; Class number________

Read carefully and then answer the questions below.

1. What are chemical reactions?
2. What does the double arrow indicate in the chemical equations?
3. On the factors that affect chemical equilibrium, answer:

    a) What happens as the temperature rises and falls in a reaction?

    b)   Effect of concentrations:

In an experiment, a solution changes from yellow to orange with the addition of a certain reagent and then back to yellow with the addition of another reagent. The following equation represents the equilibrium involved in this experiment: $2CrO_4^{2-} + 2H(aq)^{+} \rightleftharpoons 2CrO_7^{2-}$ (aq) $+ H_2 O(l)$. Knowing that chromate ions (CrO_4^{2-}) have a yellow color and dichromate ions (CrO_7^{2-}) have an orange color, answer the questions about what this means for each of the following phenomena:

    a) If the system is initially colored yellow and a few drops of hydrochloric acid (HCl) are added, this occurs:

    b) If the system is colored orange and a few drops of sodium hydroxide (NaOH) are added, this occurs:

    c) If the system is colored orange and a few drops of water ($H_2 O$) are added, this occurs:

4. What is the catalytic converter for?
5. What does Lê Chatelier's principle say?
6. When should the Lê Chatelier principle be applied?

The researcher: Manuel Pina

APPENDIX V

RESULTS OBTAINED FROM THE PRE-TESTS OF THE 10TH GRADE STUDENTS CFB.

Class. 10.1- Experimental group.

Order number	Age	Rating (0-10)
1	15	1
2	16	1
3	X	1
4	15	1
5	15	1
6	16	1
7	X	0,5
8	17	0,5
9	15	2,5
10	15	1,5
11	15	1
12	16	0,5
13	15	1,5
14	15	1,5
15	15	2
16	16	1
17	16	0,5
18	16	2
19	16	3
20	X	1
21	16	1,5
22	17	0,5
23	16	1
24	15	1
25	15	1
26	15	3,5
27	15	0,5
28	16	2,5

Class. 10.4a- Control group.

Order number	Age	Rating (0-10)
1	16	3
2	15	5,5
3	16	4
4	X	2
5	X	1,5
6	15	1,5
7	15	1,5
8	16	0
9	X	1,5
10	X	0
11	X	1,5
12	X	1
13	X	1,5
14	16	O,5
15	16	4
16	17	0
17	17	4
18	16	1,5
19	16	4
20	16	5,5
21	15	4

RESULTS OBTAINED FROM THE POST-TESTS OF 10TH GRADE STUDENTS CFB.

Class. 10.1- Experimental group.

Order number	Age	Rating (0-10)
1	16	6
2	16	5
3	15	7
4	16	0,5
5	15	8
6	15	5
7	17	6
8	15	2,5
9	15	4,5
10	16	9
11	15	1,5
12	18	5,5
13	15	2
14	16	8,5
15	15	4
16	16	3,5
17	15	5
18	15	6
19	15	7,5
20	15	9,5
21	16	5,5
22	16	6
23	16	4

Class. 10.4a- Control group.

Order number	Age	Rating (0-10)
1	16	0
2	16	1
3	16	3,5
4	16	1,5
5	16	0,5
6	17	6
7	16	3
8	15	0
9	16	1,5
10	16	2
11	15	0,5
12	17	2
13	17	4,5
14	16	4,5
15	16	4,5
16	16	2,5
17	15	1,5
18	16	3,5

EXPERIMENTS

Experiment #1.1

Lesson time: 45 minutes,

General objective: To understand the Lê Chatelier principle and the factors that affect a system in chemical equilibrium.

Topic: Chemical equilibrium - Lê Chatelier's principle

Subtheme: Factors affecting chemical equilibrium

Lesson title: Concentration effect

Objective: To demonstrate the effect of the concentration of H^+ on the chromate/dicromate balance

Materials and reagents

4- Pasteur pipettes 3- Test tubes 1- test tube holder aqueous solution of K_2CrO_4 0.1 M; aqueous solution of K_2CrO_27 0.1 M; aqueous solution of HCl 1.0 M; aqueous solution of NaOH 1.0 M.

Experimental procedure

$$2CrO2\text{-}(aq) + 2H^+(aq) \rightleftharpoons CrO72\text{-}(aq) + H_2O(l)$$

- prepare 3 test tubes, then list them.
- In tube B, add 2 mL of potassium dichromate solution (K_2CrO_27). Record the color of the solution.
- In tube A, add 2 mL of potassium chromate solution (K_2CrO_4). Record the color of the solution.
- In tube C, add 1.5 mL of potassium chromate solution. Then add 15 drops of hydrochloric acid (HCl) solution. Match the color of this solution to the color of tubes A and B.
- In the same tube C, add 3 mL of sodium hydroxide solution (NaOH) drop by drop. Homogenize the solution and match the color of tubes A and B. Consider the dilution.

Explain what happened to the chemical equilibrium after the acid and base were added.

Experiments #3.1

Activity: Concentration as one of the factors affecting chemical equilibrium

Summary: Effect of concentration

In a system, the concentrations of the reactants and products change to suit a new equilibrium, but the equilibrium constant remains the same. An example is the following equilibrium:

$$[Co(H_2O)]_6^{+2}{}_{(aq)} + 4Cl^-{}_{(aq)} \rightleftharpoons [CoCl]_4^{-2}{}_{(aq)} + 6H_2O_{(l)}$$

Pink color Blue color Cobalt II tetrachloride ions

Cobalt II hexahydrate ions

Reagents and materials: $CoCl_2$ - (cobalt chloride), HCl - 1M concentration, distilled water, 50 ml volumetric flask, spatula, 3 test tubes, analytical balance or other and gloves.

Procedures:

1. Measure 2.75g of cobalt chloride;
2. Pour the measured cobalt chloride into a 50 ml volumetric flask;
3. Add a few drops of distilled water to the volumetric flask, then shake.
4. When the cobalt chloride has dissolved, add more water up to the limit line of the volumetric flask and shake again;
5. Transfer 4 to 8 ml of the prepared solution into a test tube;
6. Add drops of HCl to the prepared solution and stir until it changes color (violet);

How can a change in concentration affect the equilibrium of a system?

To answer this question, you should transfer 1 ml of the prepared solution (violet) to conical flasks A and B, then add drops of HCl to conical flask A, shake and observe what happens. And add drops of distilled water to flask B, shake and observe what happens.

3. Stage

When you add a few drops of HCl to erlenmeyer flask A, the color immediately changes from violet to blue.

If you add a few drops of HCl, the system will increase the concentration of Cl ions⁻ . According to Lê Chatelier's principle, the system will move in the direction of consuming the Cl ions⁻ , i.e. the reaction will move in a direct direction. This will lead to the formation of products and the solution will turn from violet to blue. When you add a few drops of distilled water to erlenmeyer flask B, the color changes from violet to pink.

On the other hand, if adding a few drops of distilled water increases its concentration in the solution, the system opposes the change, causing the products to react. However, the equilibrium will shift in the direction of consuming the H $O_{2(l)}$, i.e. to the left (in the opposite direction of the reaction). This will cause the reactants to form and the solution will turn from violet to pink.

4. Stage

For erlenmeyer flask A, where a few drops of HCl have been added, if you add a few more drops of distilled water the color will change from blue to violet.

For erlenmeyer flask B, where a few drops of distilled water have been added, if more HCl is added, the color will change from pink to blue, i.e. the equilibrium will end in a blue color.

In conclusion, the given reaction obeys Lê Chatalier's principle, which establishes concentration as one of the factors that influence or affect the equilibrium of a system.

ANNEXES

PROGRAM FOR THE SECOND QUARTER 10TH GRADE, TOTAL 48 LESSONS

Tema C - Equilíbrio Químico

Objectivos gerais:
› Interpretar a ocorrência de transformações incompletas;
› Identificar qualitativamente o equilíbrio químico em sistemas homogéneos e heterogéneos;
› Compreender o princípio de Le Chatelier e as suas aplicações práticas;
› Caracterizar o equilíbrio de solubilidade em termos qualitativos.

Conteúdos:

C1 - Introdução.

C2 - Reversibilidade das transformações físicas e químicas.

C3 - Noção de Sistema.

C4 - Equilíbrio em Sistemas Homogéneos:
› Equilíbrio dinâmico;
› Estados de equilíbrio de um sistema.

C5 - Factores que afectam o estado de equilíbrio de um sistema:
› Efeito da alteração da concentração;
› Efeito da temperature;
› Princípio de Le Chantilier.

C6 - Aplicações práticas do Princípio de Le Chantilier:
› O equilíbrio químico em processos industriais;
› Sistemas biológicos;
› Sistemas geológicos.

C7 - Equilíbrio em sistemas heterogéneos:
› Equilíbrio heterogéneo;
› Equilíbrio de solubilidade.

Objectivos específicos:
› Reconhecer que existem transformações físicas ou químicas inversas;
› Ter a noção de sistema;
› Diferenciar sistemas isolados, fechados e abertos;
› Diferenciar sistemas homogéneos e heterogéneos;

13

Source: INIDE (2014).

ISCED / HUÍLA

CREDENCIAL

Para a realização de trabalho de pesquisa e para concessão de facilidades junto a **Escola Liceu Joaquim Kapango/Huambo**, credencia-se o (a) estudante **Manuel Kalelesse Pina** do 2º ano do Mestrado, Curso de Ensino da **Quimica**.

Por ser verdade e me ter sido solicitada, mandei passar a presente **CREDENCIAL**, que vai por mim assinada e autenticada com o carimbo a óleo em uso nesta Direcção.

Instituto Superior de Ciências da Educação da Huíla

Lubango, 19 de Dezembro de 2022.

O VPACPG

Bernardo Filipe Matias, PhD

GOVERNO DA PROVÍNCIA DO HUAMBO
GABINETE PROVINCIAL DA EDUCAÇÃO

DEPARTAMENTO DE EDUCAÇÃO E ENSINO

1. T. C
2. Autorizado.

001 /DEE/2023   09/1/23

AO

GABINETE DE SUA EXCELÊNCIA DIRECTOR DO GABINETE PROVINCIAL DA EDUCAÇÃO DO HUAMBO

=HUAMBO=

ASSUNTO: **Resposta à Solicitação Sobre o Estagio Pedagógico**

Melhores cumprimentos

Em resposta a Solicitação, vinda do Senhor **Manuel Kalelessa Pina**, datada de 28 de Dezembro de 2022, referente a solicitação acima referenciada, que tem como objectivo, trabalhar com os alunos da 10ª Classe do Curso de Ciências Físicas e Biológicas do Liceu Kapango – Huambo, depois de lida e analisada, somos a emitir o seguinte parecer:

Por se revestir de suma importância, a petição em referência, deve merecer a anuência favorável ao requerente pelos motivos por ele apresentados.

Atentamente.

DEPARTAMENTO DE EDUCAÇÃO E ENSINO, 05 DE JANEIRO DE 2023

O CHEFE DO DEPARTAMENTO

APOLINÁRIO SACONDO NAMALYONGO
G.P.E

"Ensinar com cientificidade e ética para humanizar a educação no Huambo"

Cidade Alta,
Praça Dr. António Agostinho Neto
Huambo
Huambo
ANGOLA

Printed by Books on Demand GmbH, Norderstedt / Germany